PLANT PROCESS OF PHARMACEUTICAL AND DYESTUFFS INTERMEDIATE

PLANT PROCESS OF PHARMACEUTICAL AND DYESTUFFS INTERMEDIATE

Prof. K.R. Desai
&
B.G. Naik

SARUP & SONS
NEW DELHI - 110002

Published by
SARUP & SONS
4740/23, Ansari Road
Darya Ganj, New Delhi - 110002
Tel.: 23281029, 41010989
e-mail : sarupandsonsin@hotmail.com

PLANT PROCESS OF PHARMACEUTICAL AND DYESTUFFS INTERMEDIATE

1st edition 2008
ISBN-81-7625-818-0

Edition 2022

PRINTED IN INDIA

Published by Prabhat Kumar Sharma for Sarup & Sons,
Laser Typesetting at Twinkle Graphics and Printed
at Mehra Offset Printers, Delhi.

Preface

Encouraged by the response to the first volume published in the year 2005 is popular not only among the organic chemist but also among those involved with practical uses of organic intermediates in various fields of science and technology.

We have written the second book is composed and written with a totally new organic intermediate should be broadly useful to chemist, chemical engineer and research students and special benefit of colleges, universities and institutions concern with the teachign of Pharmaceutical, Industrial and Applied chemistry and useful to small and large manufacturers of organic intermediates.

In "Plant Process of Pharmaceutical and Dyestuffs Intermediate" second book, we have attempted to provide a content of exceptional versatility.

This text contain totally new manufacturing process of various organic intermediate provides detailed description of the progress and completion of different chemical reaction, also giving process-control test wherever necessary, and disclosing very precise parameters that may lead to give the objective products with quality and yield in nearly close proximity to the stated values. To complete successfully in the global marketplace, organization must have the ability to produce a variety of high quality, low cost products that fully satisfy customers needs.

We are confident that as volume-1 "*Production of organic intermediate, pharmaceuticals and dyestuffs*" this book will also be widely accepted at all national and international levels of academic and technical world.

Prof. Dr. K.R. Desai
and
B.G. Naik

Contents

1

Manufacture of Meta-Nitro-Aniline

1. Principle:

One nitro-group in Meta-Dinitrobenzene molecule is selectively reduced to amino-group with sodium disulfide solution in aqueous medium. The crude product is filtered out and purified by dissolution in Hydrochloric acid and subsequent precipitation. The pure precipitated product is filtered, washed and dried in vacuum.

2. Reactions:

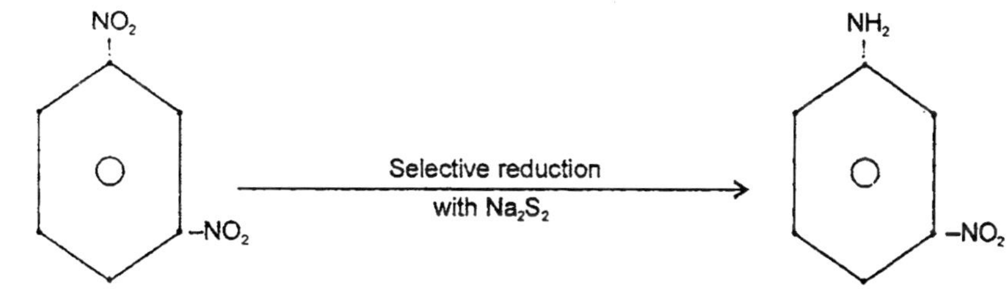

m-Dinitrobenzene m-Nitro-Aniline

3. Process-Procedure:

3.1. Preparation of Sodium disulfide

Tool A 2500 L capacity M. S., Jackettted Kettle equipped with an Anchor-type agitator driven by a 3 H. P., 3-Phase, closed Electric motor connected with a Reduction-gear to give 60 r. p.m. speed to the agitator. The Keat is further equipped with a Manhole, a Pressure-gauge, a Temperature-recorder, a Vent-line, and a bottom-outlet line leading to the Clarifier Press.

Check tool.

Charge in the Kettle

1350 L of treated water.

Start agitator and charge

421 Kgs of is 60% Sodium Sulfide lumps equivalent to

333 Kgs of real Sodium Sulfide (M. Wt. 78.0).

Circulate steam in the kettle jacket and heat the charge to 80-85° degC. Stir to solution .

Then charge gradually during about half an hour

145 Kgs of real Sulfur powder (At. Wt. 32.0).

The reaction is slightly exothermic

Stir for 1 hour at 85-87°C.

The sulfur dissolves in the Sodium Sulfide solution resulting in the formation of Sodium disulfide. Check the completion of Sodium Disulfide formation by observing presence of no undissolved Sulfur particles in a sample. If necessary, stir more for 15-20 minutes at 85-87°C.

1700 L Measure volume of the solution and adjust it to at 85-87°C by adding requisite quantity of water. Stir for 15 minutes at 85-87°C. Filter the Sodium Disulfide solution hot at 80-85°C through a Clarifier Press. Collect the clear filtrate in a suitable Vessel and hold it at 80-85°C until used.

3.2. Reduction

Tool : A 10,000 L capacity M. S., Jacketted Kettle equipped with Gate-type agitator driven by a 5 H. P., 3-Phase, closed Electric motor connected with a Reduction-gear to give 40 r. p. m. speed to the agitator. The Kettle is further provided with a Manhole, a Pressure-gauge, a Temperature-recorder, a Vent-line and a bottom-outlet line leading to the Filter Press.

Check tool.

In the clean Kettle charge

3400 L of treated water.

Start agitator.

Circulate steam in the Kettle jacket and heat to 90-92°C.

Then add

7.5 Kgs of Turkey Red Oil and

41 Kgs of real Sulfur as powder (At. Wt. 32.0).

Stir for 10 minutes a 92-93°C.

Then charge

The charge

715 Kgs of real *m*-Dinitrobenzene (M. Wt. 168.0) in the form of broken lumps.

Stir at 92-93°C until the m-dinitrobenzene dissolves. Then cool to 86-87°C, controlling the jacket steam. Now, maintaining the temperature in the range of 87-90°C, add gradually over a period of about 3 hours the clarified Sodium Disulfide solution kept ready in the above step.

When all the solution is added up, stir the batch for 30 minutes at 87-90°C.

Test on Phenolphthalein papers must be positive.

Then add atonce without any delay

65 Kgs of as is 40%w/w Caustic Soda Iye corresponding to

26 Kgs. of real Caustic Soda.

Stir 45 minutes at 87-90°C.

The reduction is complete.

Stop steam in the Kettle jacket and drain down the jacket.

Circulate cooling water in the jacket and simultaneously add about

1600 Kgs of ice to cool the batch to 32-35°C. Stir the batch for half an hour at 32-35°C.

The batch is ready for filtration.

Filter the batch through Filter Press.

Discard the M.L. to drains.

Rinse the Reduction Vessel with about

500 L of water and pass it through the Press.

Then further wash the Press cakes thrice, each time with

500 L of water.

Discard the washing to drains.

Air the Press to no drip.

Dump the Press, weigh the wet cakes and record. Weight of Wet Cakes = Approx. 625-650 Kgs. Purity this crude product as follows.

3.3 Purification:

Tool A 6000 L capacity Wooden Vat equipped with a Paddle-type wooden Agitator driven by a 5 H. P., 3-Phase, closed Electric motor connected with a Reduction-gear to give 40 r. p. m. speed to the Agitator. The Vat is provided with four baffles on the inside wall and a bottom-outlet line leading to the Pressure-Nutsche.

Charge in the clean Wooden Vat

3200 L of water. Start agitator.

		Then charge in it the total wet cakes of crude *m*-Nitro-aniline obtained in the above step.
		Stir for 1 hour to obtain a smooth slurry.
		Then charge
590	Kgs	of as is 30%w/w HCl equivalent to
177	Kgs	of real HCl.
		Stir for 15 minutes.
		Then inject direct steam into the charge and heat to 50-55°C.
		Measure volume: it will be about
4250	L	at 50-55°C.
		Stir for 2-1/2 hours at 50-55°C.
		Draw a sample of the solution and observe its clarity: the solution of MNA must be complete and clear.
		Then add
30	Kgs	of Activated Carbon and
5	Kgs	of Hyflosupercel.
		Stir for 1-1/2 hour at 50-55°C.
		Clarify the batch at 50-55°C through a Pressure-Nutsche.
		Receive the clear batch filtrate in the Isolation Vessel.
		Swill down the inner side of the Purification Vat with
250	L	of water.
		Stir and add to it
5	Kgs	of as is 30%w/w/ HCl.
		Steam the washing to 55-60°C.
		Pass this washing through the Pressure Nutsche to wash the nutsche cake.
		Collect the clarified washing in the Isolation Vessel holding the main batch filtrate. Discard the Nutsche residue to drains.

3.4 Isolation of Purified Product

Tool : a 10,000 L capacity Wooden Vat equipped with a paddle-type Wooden Agitator driven by a 7.5 H. P., 3-Phase closed Electric motor connected with a Reduction-gear to give 40 r. p. m. speed to the agitator. The Vat is provided with four baffles on the inside wall and a bottom-outlet line leading to the Filter Press.

Measure volume of the clarified batch filtrate and washing collected in this Vat from the above step. Adjust volume of the clear batch filtrate to

4600	L	with water. Start agitator.
		Add to it
5	kgs	of as is 90/92% Sodium Bisulfite.
		Stir for 5 minutes.
		Then add slowly over about 3-1/2 to 4 hours as required
1125	L	of as is 20%w/v Soda Ash solution corresponding to
225	Kgs	of real soda Ash, testing frequently the rate of fading the blue test on Congo Red papers.
		When the test on Congo Red papers is reduced to faint-blue or purple, add to the batch
4	Kgs	of as is 90/92% Sodium Bisulfite.
		Measure pH of a batch sample: it will be between 3.0-3.5.
		Then continue further the addition of 20% Soda Ash solution until the pH of the batch is adjusted in the range of 4.6-4.8.
		If necessary, add corrective quantity of extra Soda Ash solution to adjust the batch pH between 4.6-4.8.
		Then add to the batch, as required, about
800-900	Kgs	of ice to cool the batch to 25-28°C. Measure volume of the batch: it will be about at 25-28°C.
		Measure volume of the bath: it will be about
6800-7000	L	at 25-28°C.
		Then filter the precipitated product through Filter Press without any delay.
		If the filtration is delayed, the product is likely to become greenish.
		Suck the M. L. and discard to drains.
		Swill down the adhered product on the inside wall of the Isolation Vat with about
500	L	of water and pass this rinse through the Filter Press.
		Further wash the Press cakes thrice, each time with of
500	L	water and discard all washings to drains.
		Air the Press to no drip.
		Dump the wet cakes from the Press.
		Centrifuge the Press cakes to remove most
		Of the water from the product.
		Weigh the centrifuged cakes and record.
		Dry the wet centrifuged product in

Vacuum Dryer at 65-70°C.

Weigh the dry, pure product and record.

4. Yield

Weight of centrifuged wet cakes,

= Approx. 630-660 Kgs.

= Approx. 502-505 Kgs of dry, pure product containing 99.0-99.5% real *m*-Nitro-Aniline (M. Wt. 138.0)

= 500 Kgs of real *m*-Nitro-Aniline (M. Wt. 138.0)

= 85.2% of Theory on *m*-Dinitrobenzene charge.

5. Raw Materials Usage Data

Names of Raw Materials	*M. Wt.*	*Quantity Real (Kgs)*	*Moles*	*One Mole Ratio*
1. Meta-Dinitrobenzene	168.0	715.0	4.2560	1.0000
2. Sodium Sulfide, Real	78.0	333.0	4.2690	1.0030
3. Sulfur powder, Real	32.0	145.0	4.5320	1.0650
4. Turkey Red Oil	–	7.5	–	–
5. Sulfur Powder, Real	32.0	41.0	1.2810	0.3011
6. Caustic Soda, Real	40.0	26.0	0.6500	0.1527
7. HCl, Real	36.5	177.0	4.8500	1.1400
8. Sodium Bisulfite	–	9.0	–	–
9. Soda Ash	106.0	225.0	2.1220	0.4987
10. Activated Carbon	–	30.0	–	–
11. Hyflosupercel	–	5.0	–	–

2

Manufacture of p-Amino-Acetanilide

1. Principle:

p-Nitro-Acetanilide is reduced on the principle of Bechamp's Reduction with Iron and Acetic Acid at the boiling temperature in aqueous medium. The soluble iron is precipitated with Magnesium Oxide and the batch clarified. Finally, *p*-Amino-Acetanilide is isolated from the clear solution by the addition of Common nutsche cakes centrifuged to a maximum dry state. The product is evaluated and the yield determined.

2. Reactions

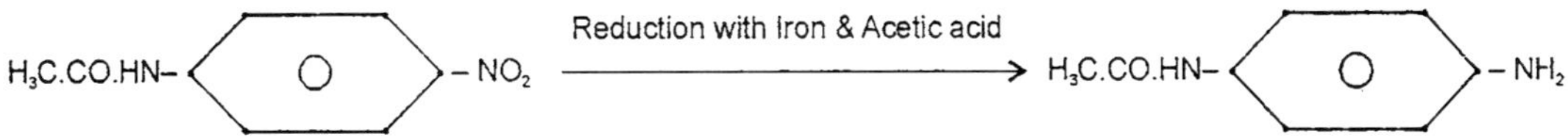

p-Nitro-Acetanilide (M.Wt. 180.0) → p-Amino-Acetanilide (M.Wt. 150.0)

3. Process-Procedure

3.1. Reduction of p-Nitro-Acetanilide

Tool A 10,000 L capacity M. S. R. L. T., jacketted Kettle equipped with a Fingure-type agitator driven by a 5 H. P., 3-Phase, closed Electric motor connected with a Reduction-gear to give 60/40 r. p. m. speeds to the agitator. The Kettle is further provided with a manhole, a direct Steam-line, a leading to the Filter Press.

Check Tool.

In the Reduction Kettle, charge

1800.0 L of water or wash-water from the previous batch.

Start agitator at a speed of 60 r.p.m.

Then charge

665.0 Kgs of Iron Filings, over and above the iron left from the previous batch.

Pass direct steam and heat the charge to 90°C.

Then add to it

50.0 Kgs of as is 85% Acetic Acid.

Stir the charge at the boiling temperature of 98-100°C and etch the iron for about ¾ to 1 hour.

Note : If HCl is used in place of Acetic acid, the reaction mass shows considerable foaming during the course of reduction.

Now reduce the speed of agitation to 40 r.p.m.

Then over a period of about 6 hours,

charge in 15-Kgs lots at a time, a total quantity of

695.0 Kgs real *p*-Nitro-Acetanilide (M.Wt. 180.0) as acid-free, centrifuged wet cakes. (About 50-60% Solid contents).

Note : The *p*-Nitro-Acetanilide used for the manufacture of *p*-Amino-Acetanilide must be practically washed neutral to Congo Red papers. Use of acidic *p*-Nitro-Acetanilide gives a bad quality of the Amino compound and also lowers the yield.

Preferably the Melting Point of a washed and dried *p*-Nitro-Acetanilide sample should come in the range of 210-222°C.

The reaction is exothermic and the reaction mass foams badly at the time of each addition of the nitro compound. During the course of entire reduction, frequently test the presence of soluble iron as per the following test:

Test : Cross-spot a drop of the reduction mass on a piece of filter paper with a drop of a clear 5% Sodium Sulfide solution. Appearance of a dark-black line at the junction of the two spots denotes the sufficiency of soluble iron.

If necessary, small corrective quantity of 85% Acetic acid may be added to maintain the soluble iron in the batch, throughout the reduction.

Continue the reduction at the reflux temperature of 98-100°C, keeping the reflux Vent line open to the atmosphere.

After the addition of Nitro compound is over, stir the batch further for 2½ hours at the reflux temperature of 98-100°C.

The foaming completely subsides at the end of the 2½ hour reduction period.

Draw about 500 ml of the batch sample in a 1 L beaker and allow it to settle.

Test the liquid for a positive test for Soluble Iron.

Clear and colourless supernatent liquor denotes the completion of reduction.

The reduction is usually complete at the end of the 2½ hour reduction period.

Measure volume of the batch and adjust it to

4800.0 L at 95-98 degC.

Note : If dried *p*-Nitro-Acetanilide is used instead of Wet Cakes, the volume of the batch at the end of reduction is correspondingly low. In this case, the volume of the batch before filtration must be adjusted to 5000 L to prevent the loss of some undissolved amine going into the filter-sludge.

3.2. Precipitation of Soluble Iron

Prepare a slurry of

34.0 Kgs of real Magnesium Oxide in

100.0 L of water in a suitable vessel.

Shake this slurry well and add it to the

Reduction mass of the batch.

Stir the batch for 10-15 minutes to enable the dissolved iron to precipitate out completely.

Cross-spot a drop of the batch sample on a piece of filter paper against a drop of 5% clear.

Sodium Sulfide solution. Development of no black or brown line at the junction of the two spots indicates that the precipitation of Soluble iron in the batch is complete.

Then add to the batch

50.0 Kgs of as is 40% w/w Sodium Bisulfite solution corresponding to

20.0 Kgs of real Sodium Bisulfite.

Stir for 5 minutes and add

12.0 Kgs of Sodium Hydrosulfite as powder.

Stir the batch for 10 minutes.

Blow the batch through the pre-heated blow-out line to the Filter Press and filter the batch hot.

Receive the clear batch filtrate into the Isolation Vessel.

Then rinse the Reduction Kettle twice, each time with

900.0 L of water.

Stir and steam each of the two washings to 90-95°C. Pass these hot washings through the Filter Press and receive both the washing filtrates alongwith the main batch filtrate in the Isolation Vessel.

Finally, take

1800.0 L of water in the Reduction kettle.

Stir and steam to 90-95°C.

Pass the washing water through the Filter Press to further wash the Iron-sludge.

Receive this clear washing filtrate in a suitable container or back into the Reduction Vessel for using as starting water in the subsequent batch. Neutralise the little alkalinity present in this washing filtrate with about 5-6 Kgs of as is 85% Acetic acid to a faint Congo test.

The iron-sludge from Filter Press is discarded.

3.3 Isolation

Tool : A 10,000 L capacity M.S.R.L. Kettle equipped with a Spade-type Agitator driven by a 5 H.P., 3-Phase, closed Electric motor connected with a Reduction-gear to give 40/20 r.p.m. speeds to the agitator. The Kettle is further equipped with a portable SS-Cooling Coil, a manhole, a Vent-line, a Pressure-gauge a temperature-recorder and a blow-out line leading to the Nutsche Filter.

Measure volume of the main batch filtrate and initial washing collected in this Kettle and adjust to

7000.0 L at 80-85°C.

Start agitator at a speed of 40 r.p.m.

Charge in it

25.0 Kgs of as is 40%w/w Sodium Bisulfite solution corresponding to

10.0 Kgs of real Sodium Bisulfite.

Stir for 10 minutes.

Then charge in it uniformly over a period of 2 hours

2100.0 Kgs of Common Salt.

Stir at 80-85°C for ½ hour after addition of the Salt.

Reduce the speed of agitation to 20 r.p.m.

Start circulation of cooling water in the Cooling Coil and cool the batch gradually to 15-18°C

over a period of 15 hours.

Stir the batch for 1 hour more at 15-18°C.

The *p*-Amino-Acetanilide is precipitated in a fine form. The pH of the batch slurry is in the range of 5.9-6.3.

Filter the precipitated product through the Nutsche Filter.

Rinse the Isolation Kettle with about

200 L of the M.L. and pass it into the Nutsche.

Measure density of the M.L. it will be about 22-23 deg Be'.

Determine the loss of *p*-Amino-Acetanilide in the M.L. with a Nitrite titre on M.Wt. 150.0; it should not show a liquor loss over 1.0-1.2% as real *p*-Amino-Acetanilide.

Note : If the pH of the M.L. is lowered from 5.9 by acidification, an extra quantity of about 12-15 Kgs of *p*-Amino-Acetanilide may be recovered. But this product is of inferior quality and also since this recovery is uneconomical, the M.L. is discarded to drains without making any recovery.

Suck the M.L. well.

Dump the Nutsche cakes.

Centrifuge the Nutsche cakes as dry as possible.

Weigh the centrifuged nutsche cakes and record.

4. Yield

Weight of centrifuged Nutsche cakes,

= Approx. 600-650 Kgs; containing 80-85%

real *p*-Amino-Acetanilide (M.Wt. 150.0).

= 500 Kgs of real *p*-Amino-Acetanilide (M.Wt. 150.0)

= 86.30% of Theory on charge of real

p-Nitro-Acetanilide.

5. Raw Materials Usage Data

	Names of Raw Materials	*M. Wt.*	*Quantity Real Kgs*
1.	*p*-Nitro-Acetanilide, Real	180.0	695.0
2.	Iron Filings	–	665.0
3.	Acetic Acid, Real	60.0	42.5
4.	Magnesium Oxide	40.0	34.0
5.	Sodium Bisulfite, Real	104.0	30.0
6.	Sodium Hydrosulfite, 80-85%	–	12.0
7.	Common Salt	–	2100.0

3

Manufacture of N:N-Bis (Hydroxy Ethyl)m-Chloro-Aniline

1. Principle:

Ethylene Oxide gas is passed in Fast Orange G Base Oil until the addition reaction is complete. The objective product solidifies on cooling to room-temperature in SS trays. The set finished product is broken to pieces, weighed and evaluated for purity and setting point. The product is finally packed in suitable containers.

2. Reactions:

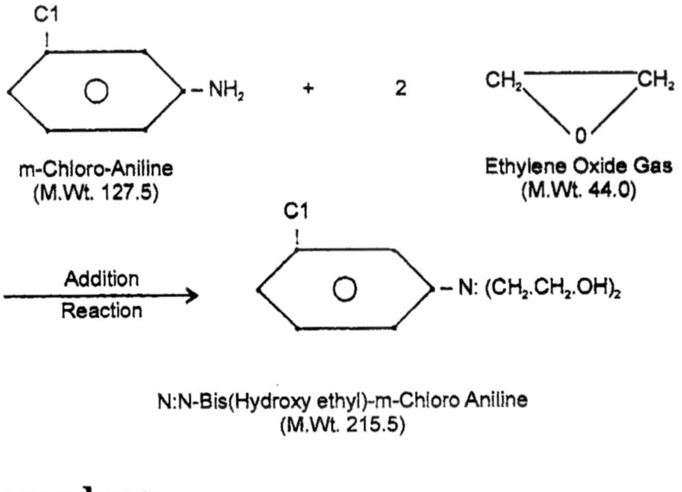

3. Process-Procedure:

3.1. Reaction with Ethylene Oxide gas

Tool : A 1000 L capacity SS or Glass-lined M. S. Kettle with an M. S. Jacket equipped with an Anchor-type agitator driven by a 3 H. P., 3-Phase, closed Flame-proof electric motor connected with a Reduction-gear to give 60 r.p.m.

speed to the agitator. The Kettle is further equipped with a charging-hole, two SS gas-feed lines reaching the anchor-base close to the shaft of the agitator, a vent line leading to the Kerosene Scrubber and bottom-outlet valve.

Check tool.

In the clean, dry Kettle, charge

300.0 Kgs of *m*-Chloro-Aniline Oil (M. Wt. 127.5).

Close manhole.

Start agitation.

Keep the Vent Valve open.

Heat the charge slowly to 85-90°C with jacket steam during about 30 minutes.

Maintain the batch temperature in the range of 85-90°C, throughout the course of reaction.

Now flush the whole system by passing Nitrogen gas for 5-7 minutes from a cylinder, through a common point on the two sets of the Ethylene Oxide gas-feed lines.

Confirm that there is no leakage in the system.

Stop the passage of Nitrogen gas.

Then slowly start the passage of Ethylene Oxide gas simultaneously from two cylinders placed vertical on the platform balance, through two different sets of gas-feed lines.

Adjust the rate of passage of the gas at 2.3 lbs pressure on the Pressure-gauge fixed on the line between the Kerosene-scrubber carbuoy and the Reaction Kettle. This pressure limit on the gauge corresponds to a passage of 2.5-3.0 Kgs of gas per hour, per cylinder.

For about first one hour, the rate of absorption of gas is slow and corresponds to about 1.5-2.0 Kgs per hour, but once the reaction is set in, the rate of absorption increases to about 2.5-3 Kgs of gas per hour, per cylinder.

As result of the reaction, the batch temperature tends to a slow rise. In this event, control the batch temperature in the range of 85-90°C by passing water for a short while through the jacket.

Control and adjust the rate of passage of gas by observing the rate of bubbling in the Kerosene oil kept in the inlet and outlet scrubber-carbuoys. Rapid passage of excessive gas beyond the limit of absorption, bubbles fast in the Kerosene oil of outlet scrubber-carbuoy and escapes out as a waste. This must be controlled and prevented during the initial stages of setting in the reaction. Subsequently, once the reaction is set in, no such difficulty will be observed and the abruption will proceed smoothly showing an average absorption rate of about 5-6 Kgs of total gas per hour.

Under ideal conditions of gas passage, no gas escapes unabsorbed as indicated by the absence of bubbling in the Koresosene oil of the vent scrubbing-carbuoy. While near the end of reaction, the rate of gas absorption again falls to about 1.5-2.0 Kgs per hour. When a gas cylinder is just about to be exhausted and the residual gas content of the cylinder is about 15-20 Kgs, just pour warm water of 40°C over the cylinder to drive out the residual gas from the cylinder for absorption.

When about

215.0 Kgs of Ethylene Oxide gas is passed, stop the gas passage and draw a batch sample.

Analyse the sample for

(i) Setting Point (About 84.5-85°C) and

(ii) TL Chromatographic test for unreacted *m*-Chloro-Aniline to match the Standard. (See Appendix I).

If the sample tests do not comply with the above, pass about

5-6 Kgs of surplus Ethylene Oxide gas at 85-90°C and retest as above.

220 Kgs About of total Ethylene Oxide gas will be necessary for the completion of reaction.

In case, if the unreacted *m*-Chloro-Aniline is more and the total gas passage corresponds to the specified quantity, stir the batch further at 85-90°C for 5 to 6 hours and retest. Completion of reaction indicates that the gas dissolved in the system has reacted on prolonged stirring. A Setting Point of 84,6-85°C indicates the completion of reaction. On completion of reaction, flush the whole system with a stream of Nitrogen gas to drive out the residual Et_2O gas collected in the empty space of the Kettle. Dump the product in SS trays through the bottom-outlet valve of the Kettle at 85-90°C and allow to cool to room temperature.

The melt solidifies in the trays.

Break the solidified product in small lumps,

Weigh and record weight.

Analyse a sample for percentage purity and Setting point. Calculate yield and record.

4. Yield:

About, 505 Kgs of as is solid product having about 99% real content (M. Wt. 215.5).

= 500 Kgs of real N:N-Bis (Hydroxy ethyl)-*m*-Chloro-Aniline (M. Wt. 215.5).

= 98.6% of Theory on charge of m-Chloro-Aniline.

5. Raw Materials Usage Data:

	Names of Raw Materials	*M. Wt.*	*Quantity Real (Kgs)*	*Moles*	*One Mole Ratio*
(i)	*m*-Chloro-Aniline	127.5	300.0	2.3529	1.0000
(ii)	Ethylene Oxide, gas	44.0	220.0	5.000	2.1250

6. Process-Notes:

(i) Ethylene Oxide is a highly inflammable gas.

It is also toxic to human life. Inhalation of the gas most be avoided.

(ii) Passage of Ethylene Oxide gas must be carefully regulated so as to allow maximum absorption in the system. This is stressed to avoid undue waste through leakage and also to prevent inhalation of polluted air by the operating personnel.

7. Specifications:

(i)	Physical Appearance	Light-yellow coloured, crystalline, hard lumps.
(ii)	Strength by total solids	99% Min.
(iii)	Setting Point	84.6°C.
(iv)	Moisture content	About 0.4%.

4

Manufacture of Aniline-2:5-Disulfonic Acid

1. Principle:

Metanilic Acid is sulfonated with about 24% Oleum at 145°C. The sulfonated mass is drowned into water, the Disulfonic Acid precipitated by salting and the product isolated by filtration.

2. Reactions:

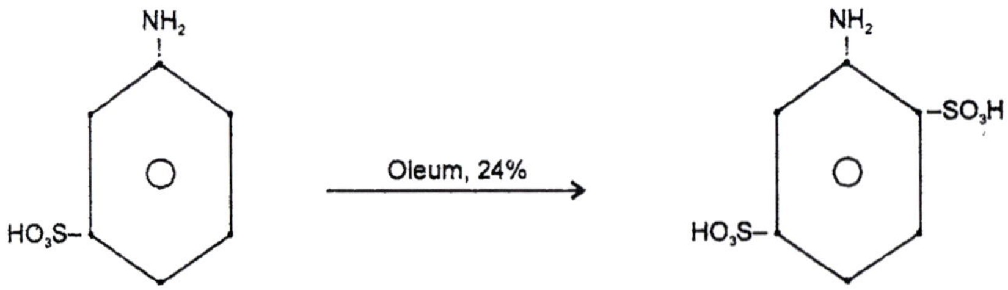

Metanilic Acid → Aniline-2:5-Disulfonic Acid

3. Process-Procedure:

3.1. Sulfonation

Tool : A 2000 L capacity C. I. jacketed Kettle equipped with an Anchor-type Agitator driven by a 5 H. P., 3-Phase, closed Electric motor connected with a Reduction-gear to give 40 r. p. m. speed to the agitator. The Kettle is further provided with a Manhole, a pressure-gauge, a Temperature-recorder and Blow-out line leading to Drawing Vessel.

Charge

870 Kgs of Monohyrate (100% Sulfuric Acid).

Start agitator at 60 r. p. m. speed.

Then charge

365 Kgs of Real Metanilic Acid (M. Wt. 173.2) as a dry powder evenly over a period of 2 hours allowing the temp. no to exceed 45°C.

Finally, charge

525 Kgs of as is 65% Oleum during about 1 hour holding the temperature below 60°C.

Start a passage of steam in the jacket and heat he sulfonation to 125°C during about 45 minutes.

Turn off steam passage at this stage and allow the temperature to rise to 145 = 1°C as a result of the Heat of Reaction. (See note below).

Stir the Sulfonation at 145 ± 1°C for 8 hours by circulating a controlled steam flow in the jacket. At the termination of 8 hours period of sulfonation, drain off steam from the jacket and start circulation of ordinary water to cool down the Sulfonation mass of 100°C.

Draw a sample of the Sulfonation mass at 100°C for checking the completion of sulfonation as described below.

Drain off water from the jacket and restart the passage of steam to heat up the sulfonation back to 145 ± 1°C and continue sulfonation at this temperature while the sample is being tested for completion of sulfonation.

Note : If the sulfonation temperature doesn't reach to 145°C of its own as a result of the Heat of Reaction, apply steam in the jacket to heat to this limit. In case, due to low steam pressure the said temperature is not obtainable, then continue the sulfonation reaction at 125-130°C for 12 hours.

Test for Completion of Sulfonation

Diazotisation

Place 45 gms of crushed ice in a 600 ml beaker, surrounded externally by ice. Add to it slowly 45 gms of the sulfonation sample under stirring with a rod. Then add to it under mechnical agitation, 70 ml of 30%w/w Caustic Soda solution so that the mixture still remains Congo acid. Then add 20 ml of 30 HCl and cool the solution externally to 10-12°C. Diazotise the mixture by adding slowly about 25 ml of a 2N Sodium Nitrite solution from a burette to a 3-minute end point on KI-Starch paper.

Dilute the diazo solution with cold water to 10 times the volume of the 2N Sodium Nitrite solution required for diazotization. The volume of the final diluted Diazo will be about 250 ml at 10-12°C.

Coupling

Dissolve 7.40 gms of pure Beta-Naphthol in an 800 ml beaker containing 200 ml of water and 15 ml of 30%w/w Caustic Soda solution by stirring mechanically.

Add to it 5 gms of Soda Ash powder and stir to solution.

Finally, to the Beta-Naphthol solution, add quantitatively the diazo solution prepared above during about 5-6 minutes under continues agitation.

The diazo couples fast with the Beta Naphthol in solution.

Stir the Coupling for 5 minutes without temperature control and take a Chromatogram as explained below.

Chromatogram

Place one drop of the Coupling solution under agitation on a sheet of Whatman No. 2 Filter Paper. Diffuse the spot by very gradual dropping of distilled water from a burette fixed over it until the circle diffuses to a diameter of about 5-6 Cms. Dry the chromatogram rapidly in an incubator.

Evaluation of Chromatogram

The Chromatogram shows the disulfo-and trisulfo-acids as orange-red and light-brown outer rings respectively and any unchanged Metanilic Acid as a pink inner ring. The sulfonation is complete when no inner ring is evident or when the latter is only faint-pink coloured. Presence of no inner pink ring at all indicates an ideal completion of sulfonation.

If the sulfonation is found to be partially incomplete, stir the batch further for 1-1/2 hours at 145 ± 1°C and repeat the above test. The sulfonation will be practically complete by now. When the sulfonation is complete, stop heating and cool the batch to 100°C by circulating water in the jacket.

3.2. Drowning and Isolation

Tool : A 4000 L capacity M. S. Rubber-lined and Brick-lined Vessel equipped with cooling coil and Gate-type Agitator with a Monel Paddle driven by a 7.5 H. P., 3-Phase, closed Electric motor connected with a Reduction-gear to give 25 r. p. m. speed to the agitator. The Vessel is further provided with a temperature-recorder, and a steam pipe.

		Charge
2000	L	of Treated water.
		Start agitator.
		Blow in the sulfonated mass from the Sulfonator slowly during about 30 minutes, the temperature of the drowning mixture rising to 85-90°C. Blow a passage of live steam in the drowned mass and heat up to 96-98°C. Maintain this temperature for 1 hour under agitation.

This helps the 2:4:5-Trisulfonic Acid to Hydrolyse back to 2:5-Disulfonic Acid.

Stop steam.

Measure Volume of the batch: it will be about

2800-2850 L at 96-98°C.

Then charge to the batch, evenly during about 1-1/4 hours,

285 Kgs of Common salt so as to serve a Salt concentration of 10% by volume in the system. The temperature may fall to about 80°C.

Stop heating.

Stir the batch without temperature control until it cools to 60°C.

The circulate ordinary water in the cooling coils and cool the batch further to 35-40°C. Finally, circulate chilled water in the coils until the temperature falls to 16-17°C.

Stir the batch for 1 hour at 16-17°C.

Measure the batch volume and record.

Draw a batch sample and determine the

Liquor-loss as described below.

Determination of Liquor-loss

Filter 500 ml of the measured batch sample through a dry Buchner funnel. Press the Filter cake, but do not wash. Measure the volume of the filtrate and record. Measure 100 ml of the clear filtrate and place it in a 1.5 L beaker containing 1000 ml of cold water of 10°C. Place under mechanical stirring and add 25 ml of Conc. HCl. Titrate the solution at 10-12°C with a Standard N/2 Sodium Nitrite solution to a 3-minute and point on KI-Starch paper.

Let,

V = Volume of the batch in Litres,

v = Volume of the filtrate in ml of the 500 ml sample taken,

T = ml. Burette titre of N/2 Sodium Nitrite.

Then,

$$\frac{T \times \text{N. factor of N/2 NaNO}_2 \times v \times V \times 9528}{10^{10}}$$

= Liquor loss as % on Theory.

Note : The Liquor-loss in the total batch M. L. does not exceed 6% of Theory.

If the Liquor-loss is in excess of 6%, add extra Salt at a rate of 1% by volume and stir for 1-1/2 hours at 16-17°C and retest. Usually, more salting than extra 1% is not needed.

3.3. Filtration

Tool : One M. S. Brick-lined Nutsche Filter having 7.4 Sq. Meters basket area.

Filter the batch at 16-17°C.

Suck the M. L. Compress the pasty product

Well and apply vacuum for 20 hours.

Finally, wash the Nutsche cake with

300 L of 15%w/v Common Salt solution and suck further for 20 hours. Press and suck the cake very well.

Discharge the pasty wet cakes into tared casks, weigh the wet product and record.

Analyse a representative sample of the wet product for real content and calculate yield.

4. Yield

Weight of wet product,

= Approx. 1100-1200 Kgs.

= About 500 Kgs of Real Aniline-2:5-Disulfonic Acid (M. Wt. 253).

= 94% of Theory on Metanilic Acid charge.

5. Raw Materials Usage Data

	Name of Raw material.	*As is Kgs.*	*%age.*	*Real Kgs.*	*M. Wt.*	*Moles*	*Mole Ratio*
1.	Metanilic Acid, Dry	372.5	98.0	365.0	173.2	2.1080	1.0000
2.	H_2So_4 –Monohydrate	870.0	100.0	870.0	98.0	9.6450	4.2120
3.	Oleum, 65%	525.0	65.0	341.2	80.0	4.2650	2.0230
4.	Common Salt (Isolation)	285.0	–	–	–	–	–
5.	Common Salt (Washing)	45.0	–	–	–	–	–

5

Manufacture of p-Nitro-Phenol

1. Principle:

p-Chloro-Nitrobenzene is reacted with Caustic Soda under pressure in an autoclave whereby the chloro-group gets hydrolysed to Hydroxy-group. The Sodium Salt of *p*-Nitrophenol is isolated by crystallization wherefrom free *p*-Nitro-Phenol is obtained by treating its solution with HCl.

2. Reactions

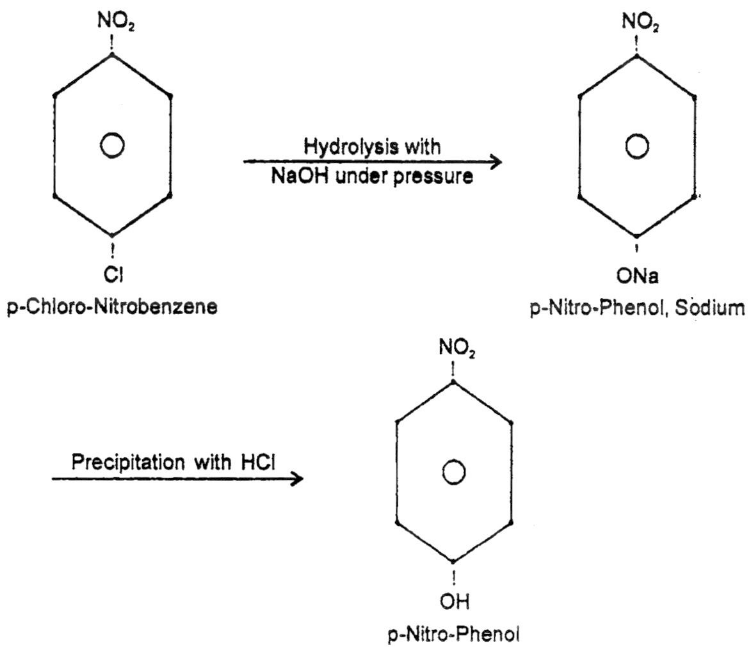

3. Process-Procedure:

3.1. Hydrolysis

Tool : A 3000 L capacity jacketed, C. I. or M. S. Autoclave designed for 150 p. s. i. Working Pressure. It is equipped with an anchor-type agitator driven by a 5 H. P., 3-Phase, closed Electric motor connected with a Reduction-gear to give 20 r. p. m. speed to the agitator. The autoclave is further provided with a Temperature-recorder, a Pressure-gauge, a Distillation unit with a Condensor and a graduated Receiver and a blow-out line leading to the Crystalliser.

In the clean Autoclave.

Charge

1325 L of water.

Start agitator.

Then charge

530 Kgs as is 40%w/w Caustic Soda lye corresponding to

212 Kgs of real Caustic Soda followed by

333 Kgs of real p-Chloro-Nitrobenzene (M. Wt. 157.5).

Note : In the first batch of a campaign, charge 1 kg of *p*-Nitro-Phenol to the batch).

Close the Autoclave.

Check up that the Valves on blow-out line and

Distillation column are closed.

Pass steam in the jacket and gradually heat the batch to 165-170°C during about 1-1/2 hours. During this heating, record temperatures and the corresponding pressures at every interval of 15 minutes on the Log-sheet.

The pressure reached at the temperature of 165°C will be about 75-80 p. s. i. In any case, the pressure of 75-80 p. s. i. is the ideal condition for the hydrolysis reaction and that, the reaction must be carried out within this pressure range.

Stir the batch at the said temperature and pressure for 5 hours.

The Hydrolysis is complete.

Then stop steam and cool the batch to 95°C.

Open the autoclave and record volume.

Adjust the Volume to

2475 L at 95-100°C with water.

Close the autoclave.

Open the valve on the Distillation column and start circulation of water in the Condenser. Then pass steam in the autoclave jacket and start distillation of water at a steady rate. Distill out

800 L of water from the batch so that the end distillate is clear and colourless.

Then stop further distillation.

The volume of the batch will be about 1675 Litres.

Close the valve on the Distillation column and open the one on the blow-out line leading to the Crystalliser.

Blow air into the autoclave and transfer the batch mass to the S. S. Crystalliser.

3.2. Crystallisation & Isolation of Sod. Salt

Measure volume and temperature of batch in the Crystalliser.

Start agitator of the Crystalliser at a high speed of 20 r. p. m.

The batch mass already contains about 124 Kgs of common salt formed as a result of the Hydrolysis reaction.

Then add to it enough extra Vacuum Salt so as to serve a total Salt concentration of 28% by volume in the batch. This requires about

350 Kgs of Vacuum Salt.

Heat the batch of 100°C by passing steam through the jacket of Crystalliser. Then stop steam.

Reduce the agitation speed to 10 r. p. m. and stir to cool to 80°C.

Finally, circulate cooling water through the Crystalliser jacket and slowly cool the batch to 30°C over a period of about 5 hours.

Filter the product through a Centrifuge.

Dump the wet cakes of Sodium Salt from the Centrifuge, weigh and record.

Weight of centrifuged Sodium Salt cakes,

= Approx. 500-520 Kgs.

3.3. Precipitation of Free Phenol:

In a suitable Vessel,

Charge

2000 L of water.

Start agitator and charge the Sodium Salt wet cakes of *p*-Nitro-Phenol obtained under the previous step. Heat if necessary to 45-50°C and stir to solution.

Cool the solution to 22-25°C by external cooling. Then add, as required, about

250 Kgs of as is 30%w/w HCl corresponding to

75 Kgs of real HCl to obtain a just faint-purple test on Congo Red papers.

Stir for 2 hours holding the temperature between 22-25°C.

Filter the precipitated product through a Centrifuge.

Wash the wet product with ice-cold water to remove chlorides.

Dump the centrifuged cakes, weigh and record. Analyse a representative sample of the centrifuged cakes and calculate yield.

Pack the wet product in steel drums having polythene in liner bags.

4. Yield:

Weight of centrifuged product cakes,

= Approx. 325-350 Kgs.

= 250 Kgs of real *p*-Nitro-Phenol (M. Wt. 139.0)

= 85% of Theory on charge of *p*-Chloro-Nitrobenzene.

5. Raw Materials Usage Data:

	Name of Raw Materials	*M. Wt.*	*Quantity Real (Kgs)*	*Moles*	*One Mole Ratio.*
1.	*p*-Chloro-Nitrobenzene	157.5	333.0	2.1130	1.0000
2.	Caustic Soda	40.0	212.0	5.3000	2.5080
3.	HCl, real	36.5	75.0	2.0550	0.9727
4.	Vacuum Salt	–	350.0	–	–

Caution : Sodium Salts of Nitro-Phenols are explosive and enough care should be taken that they do not get dried up at any stage of reaction.

6

Manufacture of p-Chloro-o-Nitro-Phenol

1. Principle:

2:5-Dichlro-Nitrobenzene is reacted with Caustic Soda under pressure in an Autoclav whereby the Chlorine atom in position 2 gets hydrolysed to Hydroxy group resulting in the formation of the objective compound. A certain volume of water is then distilled off from the system and the batch is finally filtered, washed and the cakes centrifuged to raise the solid contents of the wet product. The wet centrifuged product is finally evaluated for purity and used as Sodium Salt without drying.

2. Reactions:

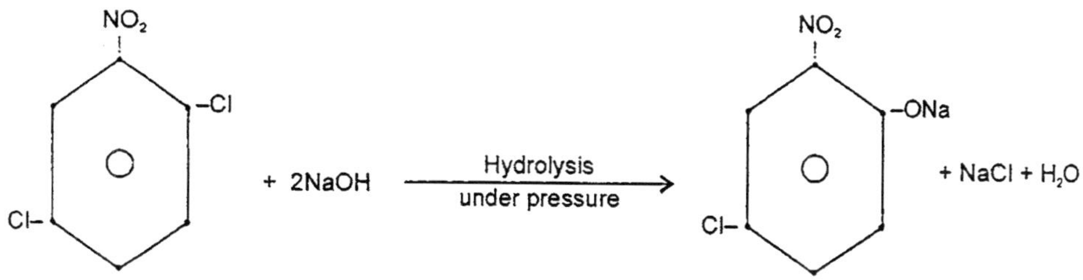

2:5-Dichloro-Nitrobenzene (M.Wt. 192.0)

p-Chloro-o-Nitro-Phenol Sodium Salt (Free Phenol, M.Wt. 173.5)

3. Process-Procedure:

3.1. Hydrolysis of 2:5-DCNB

Tool : A 7,000 L capacity Special M. S. autoclave with a jacket for injection of 20 p.s. i. steam and tested for 100 p. s. i. Working Pressure. The Autoclave is equipped with a Special M. S. Anchor-type Agitator driven by a 15 H. P., 3-Phase, flame-proof electric motor connected with a Reduction-gear to give 20

r. p. m. speed to the agitator. The Autoclave is further equipped with a Temperature-Regulator-cum-Control, a Pressure-gauge, a Manhole, a liquid-charging line, a jacketed Blow-out line leading to the Distillation kettle and a Bottom-discharge valve.

		Check tool.
		In the clean Autoclave,
		Charge
4,000.0	L	of treated water.
		Start agitator and charge
615.0	Kgs	of real 2:5-Dichloro-Nitrobenzene (M. Wt. 192.0) in the form of small, crushed lumps.
		Measure volume and adjust it to
5,000.0	L	with more of treated water as required.
		Then charge
805.0	Kgs	of as is 40%w/w Caustic Soda lye corresponding to
322.0	Kgs	of realm Caustic Soda followed by
6.0	Kgs	of as is 25% Ammonia Liquor.

Close manhole and all the valves on the Autoclave. Inject 15 p.s.i. steam in the Autoclave jacket and slowly heat the charge to 135°C during about one hour.

Make sure that there is no leakage in the autoclave. Stop further passage of steam in the jacket at 135°C and observe the rise in temperature as a result of exothermic heat.

The temperature generally rises to 145°C of its own. If not, carefully heat to 145°C with a controlled passage of steam through the jacket.

The pressure attained at 145°C is 50 p.s.i.

Then carefully stir the hydrolysis at 145°C under 50 p.s.i. built up pressure for a period of 4-1/2 hours. The hydrolysis is positively complete by now.

Stop steam.

Drain down steam from the jacket and stir to cool to 90°C.

The down steam from the jacket and stir to cool to 90°C.

The batch is now ready for transfer to the distillation Kettle.

3.2. Distillation of Water

Tool : A 7000 L capacity, jacketted, M. S. Distillation Kettle with an M. S. Anchor-type Agitator driven by a 15 H. P., 3-phase, Electric motor connected with a Reduction-gear to give 15 r.p.m. speed to the agitator. The Kettle is further equipped with a Thermometer, a Pressure-gauge, a liquid charging line, a Vent line, observation glasses, a Manhole, a Bottom-discharge line leading to the Nutsche Filter and a complete Distillation Unit consisting of a shell and Tube Condensor and a graduated Distillate Receiver.

Check tool.

In the clean Distillation Kettle, receive the hydrolysed batch from the Autoclave, through the jacketed. Blow-out line which is previously heated with steam.

Start agitator.

Open the valve on the Distillation line and circulate water in the Condensor jacket.

Blow steam in the jacket of the Distillation Kettle to heat the batch to 95-110°C.

Distill water from the batch at this temperature till about

2,200.0 L of water is distilled out from the batch.

Distillation time is about 10-12 hours.

At the end of distillation of the specified volume of water, stop heating and drain down steam from the Kettle jacket.

Circulate ordinary water at 30°C in the jacket to cool the batch to 60°C. Then circulate chilled water in the jacket to cool the batch further to 15°C.

Total cooling period = About 4 hours.

The batch is now ready for filtration.

Blow the batch to the Nutsche Filter.

Suck the M. L. well and release it to drains.

Wash the Nutsche cakes with only

200.0 L of ice-cold water, under slow leeching of the cakes under suction.

Suck the cakes well.

Dump the Nutsche Filter.

Weight the Nutsche cakes and record.

Centrifuge the nutsche cakes to raise the solid contents to about 75%.

Dump the centrifuge, weigh the wet cakes and record. Draw a sample of the wet centrifuged product and determine its purity as described in the attested process. The wet centrifuged product is used as such without drying.

Record Yield.

4. Yield

Weight of centrifuged cakes,

= Approx. 665-700 Kgs.

= 500 Kgs of real p-Chloro-o-Nitro Phenol (M. Wt. 173.5).

= 90.0% of Theory on the charge of real

2:5-Dichloro-Nitrobenzene.

5. Raw Materials Usage Data

	Names of Raw Materials	*M. Wt.*	*Quantity Real (Kgs)*	*Moles*	*One Mole Ratio*
(i)	2:5-Dichloro-Nitrobenzene	192.0	615.0	3.2031	1.0000
(ii)	Caustic Soda	40.0	322.0	8.0500	2.5132
(iii)	Ammonia liquor, 25%	–	6.0	–	

Method of Analysis for p-Chloro-o-Nitro Phenol M. Wt. 173.5

Assay

Weigh accurately 20. gms of the material and transfer to a litre beaker with about 200 ml water. Dissolve by adding enough 10% w/v sodium hydroxide solution until it is distinctly alkaline to phenolphthalein test papers. Transfer the solution to a 1000 ml volumetric flask and dilute to the mark with water. Mix well.

From this solution, pipette 100 ml into a wide-mouthed conical flask and add 20 gms of zinc dust in the flask. Close the flask with a rubber stopper through which a reflux air condesor and a long-stem separating funnel are inserted. Add drop wise, with thorough shaking 100 ml of concentrated Hydrochloric acid from the separating funnel. When the reaction subsides, place the flask on the Bunsen flame and reflux gently for 30 minutes. Then wash the inner sides on the air Condensor. Filter hot and wash with 100 ml hot water. Cool the filtrate to 0.5°C with washed ice. Then add 5 gm of Potassium Bromide and stir mechanically. Titrate with a Standard N/10 Sodium Nitrite solution until an immediate blue test on Starch-Iodide test papers is produced that can be obtained repeatedly during a period of 5 minutes without further addition of the Nitrite solution. Conduct a Blank titration in the same manner using the reagents in the same proportions and under identical conditions.

Calculations

Ml N/10 $NaNO_2$ Burette Reading-Ml Blank Reading.

= Nett Ml N/10 $NaNO_2$ used by the reduced material.

Then.

Nett Ml N/10 $NaNO_2 \times 0.01735 \times 1000/100 \times 100/20$

= % Nitrite Value as

p-Chloro-o-Nitrophenol (M. Wt. 173.5)

7

Manufacture of Ortho-Anisidine

1. Principle:

o-Nitro-Chloro-Benzene is reacted with sodium Methoxylate solution in an Autoclave to convert it to o-Nitro-Anisole. This oily material is isolated, washed and subjected to Iron-Formic Acid reduction to convert it to o-Anisidine. This is separated from the batch by distillation and the crude o-Anisidine Oil so obtained is finally purified by vacuum distillation.

2. Reactions:

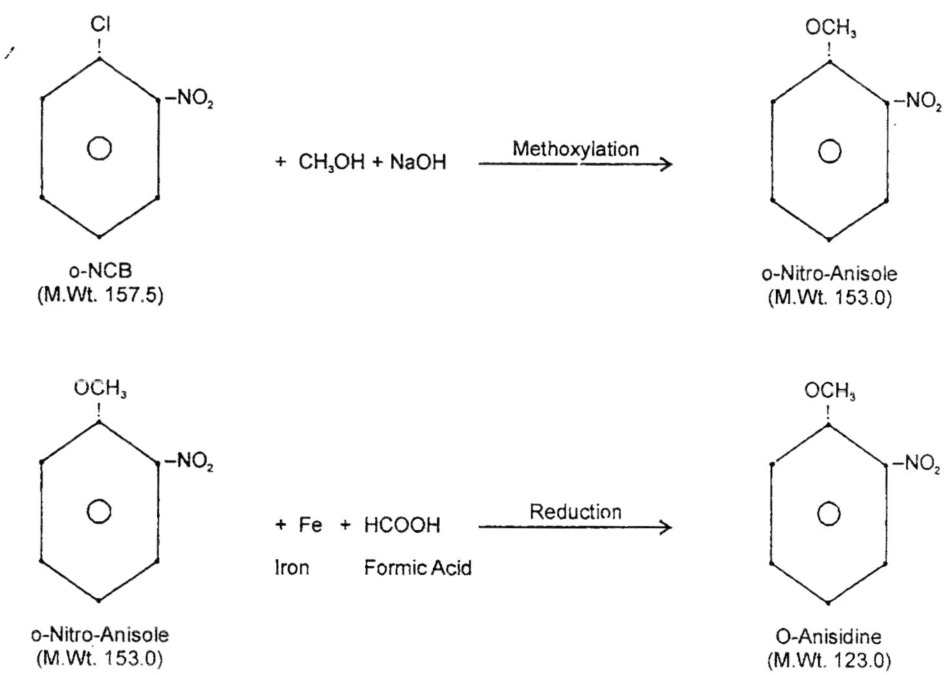

3. Process-Procedure:

3.1. Sodium Methoxylate solution

Tool A 1500 L capacity M. S. or S. S. Kettle with an M. S. jacket and equipped with an M. S./S.S anchor-type Agitator driven by a 3 H. P., closed Flame-proof electric motor connected with a Reduction-gear to give 40 r. p. m. speed to the Agitator. The Kettle is further provided with a total Reflux system, a manhole, a liquid charging line, a temperature-recorder, a pressure-gauge, a blow-out line leading to the Methoxylation-autoclave and a bottom-discharge valve.

Check tool.

In the clean and dry Reaction Kettle,

Charge

550.00 L of fresh Methanol.

Start agitator and charge gradually over a period of about 2 hours

112.00 Kgs of real Caustic Soda as flakes, maintaining the reaction temperature between 35 & 40°C by circulating ordinary water in the Kettle-jacket as may be necessary.

The reaction between Caustic Soda and Methanol being exothermic, the temperature has to be checked and maintained in the specified limit. Measure volume and adjust to

600.00 L by charging requisite quantity of fresh Methanol. Stir for half-an-hour at 35-40°C.

Analyse a sample of this Sodium Methoxylate solution for real Caustic Soda content. The NaOH content should be about 18.6% 19.0% W/Vol; if not adjust the NaOH content of the solution in the range of 18.6-19.6% w/vol.

The total, real Caustic Soda content of the entire solution should be about 111.5-112.5 Kgs.

The final volume of the Sodium Methoxylate solution is about

600.0 L at 35-40°C.

This solution is now ready for use in the methoxylation of o-NCB.

3.2. Methoxylation of o-NCB

Tool A 2,000 L capacity M. S. Autoclave designed for giving 100 p. s. i. Working Pressure an M. S. jacket and equipped with an M. S. Anchor-type Agitator driven by a 5 H. P., closed Flame-proof electric motor connected with a Reduction-gear to give 30 r. p. m. speed to the agitator. The Autoclave is further equipped with a temperature-recorder, a pressure-gauge, observation glasses, a liquid charging line, a manhole, a bottom-discharge valve and a blow-out line leading to the Conical Vessel for Isolation.

Check tool.

In the dry, clean Autoclave,

Charge

110.00 L Methanol (fresh), start stirring and suck into it

250.00 Kgs of real, previously molten o-Nitro-Chloro-Benzene (M. Wt. 157.5) from the Melting pots.

Measure volume; it will be about

325.00 L Now receive into this the Sodium Methoxylate solution maintained at 35-40°C in the above step, gradually during about one hour and controlling the temperature of the charge at about 45-50°C.

Close the Autoclave and slowly during about two hours heat the charge to 100°C.

Check up the Autoclave is well sealed and there is no leakage.

Note pressure and temperature of the Autoclave charge.

The temperature will be 100°C an the pressure will be in the range of 60-65 p.s.i.

Stir the batch under this temperature and pressure for 9 to 10 hours.

The methoxylation of o-NCB is positively complete under these conditions.

Draw a batch sample, isolate o-Nitro-Anisole and take its Setting point. If it is 10°C, the reaction is complete.

Cool the batch to 40°C by circulating cooling water in the Autoclave jacket.

Measure volume of the methoxylated batch charge; it will be about

1, 200.00 L at 40°C.

The batch is now ready for isolation.

Isolate the batch as described in the succeeding step.

3.3 Isolation of o-Nitro-Anisole

Tool A 7,500 L capacity SS, closed, Conical Vessel (M. S. Vessel can do) with an M. S. jacket and equipped with an SS Agitator driven by a 5 H. P., closed, Flame-proof electric motor connected with a Reduction-gear to give 40/30 r. p. m. speed to the Agitator. The Vessel is further equipped with a Thermometer, a liquid charging line, a manhole and conical bottom-discharge with observation system.

Check tool.

In the clean Conical Isolation Vessel, measure

4,900.00 L of ordinary water.

Start agitator and receive into this the methoxylated bath charge from the above step, slowly during about one hour, maintaining the drowning temperature below 50°C by circulating ordinary or cold water in the Vessel-Jacket, as is necessary.

Measure volume, temperature and pH of a batch sample.

Volume	:	About 6,500 litres.
Temperature	:	Below 42-45°C.
pH	:	10.5 to 11.0

Then while stirring, add to the batch slowly during one hour. As required, about

125-130 Kgs of as is 30% Hydrochloric Acid to adjust pH of the batch in the range of 6.5 to 7.0

Stir the batch for 15 minutes and re-check the pH; it must be between 6.5-7.0.

Circulate chilled water in the Vessel-jacket and cool the batch charge to 25°C.

Stir the batch for one hour at 25°C.

Stop stirring and allow the batch to settle for two hours .

On settling, the batch liquid separates out two distinct layers; the lower layer of oily o-Nitro-Anisole and the upper one of Methanol-water mixture. The oily layer of o-Nitro-Anisole is separated and collected in drums.

The separated o-Nitro-Anisole is further washed with water as described in the following step.

The aqueous methanol layer is pumped to the Storage tank of Methanol Rectification Unit for methanol recovery. The recovery of Methanol from this aqueous mixture is about 75 to 80% of the total Methanol charge.

3.4. Washing of o-Nitro-Anisole Oil

Charge the o-Nitro-Anisole Oil collected in drums under the above step in the same empty Conical Separating Vessel.

Charge in it

1.300.00 L ordinary water and stir for half hour.

Stop stirring and allow the batch to settle for one hour to separate into two layers.

Collect the washed o-Nitro-Anisole oil into clean drums. Use this washed o-Nitro-Anisole oil for reduction as described under the succeeding step.

Discard the aqueous layer to drains.

Record yield of the crude, washed o-Nitro-Anisole Oil as follows.

Wt. of the washed, crude, as is

o-Nitro-Anisole Oil,

= Approx. 225-230 Kgs, containing about

92-93% real o-Nitro-Anisole (M. Wt. 153.0).

= About 205.0 Kgs real o-Nitro-Anisol (M. Wt. 153.0).

3.5. Reduction of o-Nitro-Anisole

(of Two Batches)

Note Since the reduction involves smaller Working Volume, washes o-Nitro-Anisole Oil from two batches each of 250 Kgs of o-NCB size, is together subjected to reduction.

Tool A 2,000 L capacity C. I. kettle with an M. S. Jacket equipped with an M. S., Fingure-type Agitator driven by a 5 H. P., closed electric motor connected with a Reduction-gear to give 40 r. p. m. speed to the agitator. The Kettle is further equipped with a Thermometer, a liquid charging line a direct steam line, a manhole Reflux system, distillation system consisting of a Shell and Tube Condensor, a Separator and a bottom-discharge valve.

Check tool.

In the clean Reduction Kettle, charge

300.00 L of ordinary water.

Start agitator and charge

510.00 Kgs of Iron Filings.

Stir for 15 minutes and charge

8.00 Kgs of real Formic Acid (M. Wt. 46.0).

Close manhole and heat the charge to 100°C with jacket steam during about one hour.

Reflux starts at 100°C.

At the reflux temperature of 100°C. Slowly charge over a period of two hours

225-230 Kgs of as is 90-92% washed o-Nitro-Anisole Oil corresponding to

205.00 Kgs of real o-Nitro-Anisole (M. Wt. 153.0).

N. B. : This quantity of o-Nitro-Anisole is obtained from one batch of 250.0 Kgs of real o-NCB. At the end of addition of Oil. Stop steam in the Kettle-Jacket and stir the batch for 15 minutes.

Then charge

4.5	Kgs	of real Formic Acid (M. Wt. 46.0).
		Stir for 15 minutes.
		Again start stem and heat to the reflux temperature i. e. 100°C.
		At this reflux temperature of 100°C, slowly charge over a period of two hours, a second lot of
225-230	Kgs	of as is 90-92% washed o-Nitro-Anisole Oil corresponding to
205.00	Kgs	of real o-Nitro-Anisole Oil (M. Wt. 153.0), obtained from the second 250.0 Kgs o-NCB batch size.
		At the end of addition of the second lot of o-Nitro-Anisole Oil, stop steam in the Kettle-jacket and stir the batch for 15 minutes.
		Then charge
4.5	Kgs	of real Formic Acid (M. Wt. 46.0).
		Stir the batch for 15 minutes.
		Again start steam in the Kettle-jacket and heat the batch to the reflux temperature.
		Stir the batch under reflux at 100°C for five hours. Stop steam, cool to 90°C, draw a batch sample and test for completion of reduction (Zinc-Acetic acid reduction and the differential Nitrite titres).
		Also isolate o-Anisidine from a small batch sample and determine its setting Point. The Setting Point must be 5.5°C. Usually, the reduction is complete by now but in case, if it is found to be partially incomplete, stir the batch under reflux at 100°C for two hours more and re-check the reduction test.
		The reduction is complete.
		Measure volume of the batch the adjust it to
1,100.0	L	at 90°C with water.
		At about 90°C, add to the batch,
		as required, about.
20-22	Kgs	of Magnesium Oxide (M. Wt. 40.0) to obtain strong Orange-red test on B. Y. papers. This neutralizes the acidity of the batch.

Distillation of o-Anisidine

Close the manhole of the Kettle and heat the batch to the reflux temperature of 100°C with jacket steam. Reflux the batch at 100°C for 15 minutes.

Now close the reflux and open the valve on the Distillation-line. Start circulation of ordinary water in the jacket of the descent Condensor.

Start distillation of o-Anisidine Oil and Water mixture.

Collect the distilled o-Anisidine Oil, through

Separator, in a clean drum.

Return the separated water back to the Reduction Kettle.

Total Distillation time = About 36 to 40 hours.

At the end of completion of distillation filter the residual batch charge through Filter Press.

Check the Press sludge for loss of o-Anisidine and discard.

Weight of crude, distilled o-Anisidine

Obtained = Approx. 305-310 Kgs.

This yield is obtained from a total quantity of 500 Kgs of real o-Nitro-Chloro-Benzene, i. e. from two batches, each of 250 Kgs, or real o-NCB.

Vacuum-distil this crude o-Anisidine Oil as described in the following step to obtain pure o-anisidine.

3.6. Vacuum Distillation of crude o-Anisidine:

Tool A 750 L capacity SS Vacuum Distillation Kettle having an M. S. Jacket designed for feeding 100 p.s.i. steam. The Distillation Still is further equipped with a Jacketted-condensor and a Receiver, a Liquid charging Line, a temperature-recorder and connected to a powerful Vacuum line.

Check tool.

In the clean, dry, Vacuum Distillation Still, charge about

305-310 Kgs of as is crude o-Anisidine Oil obtained through distillation under the previous step.

Note : This quantity of crude *o*-Anisidine Oil is obtained from the total quantity of 500 Kgs of real *o*-NCB, as declared earlier.

By injecting steam in the still-jacket, heat the Oil to 100°C and distil out water to dehydrate the *o*-Anisidine Oil under atmospheric pressure.

Then close and seal the system.

Start Vacuum and gradually take full vacuum of 30 inches.

Heat to 125°C and slowly distil the Oil.

Distillation will be compete at 150°C.

Fill the pure, vacuum-distilled o-Anisidine Oil in clean drums from the Receiver.

Weight of pure, vacuum–distilled *o*-Anisidine Oil, = Approx. 266-275 Kgs.

Record yield.

4. Yield

Weight of crude, distilled, *o*-Anisidine Oil,

= Approx. 305-310 Kgs.

Weight of pure, vacuum-distilled o-Anisidine Oil,

= Approx. 266-275 Kgs, having about 99.0-99. 5% purity (M. Wt. 123.0).

= Average 270.0 Kgs of real *o*-Anisidine (M. Wt. 123.0).

= 69.0% of Theory on the charge of 500.0 Kgs of real o-Nitro-Chloro-Benzene (M. Wt. 157.5).

This corresponds to a yield.

= 135 0 Kgs of real o-Anisidine (M. Wt. 123.0) obtained from the initial batch size of 250.0 Kgs of real o-Nitro-Chloro-Benzene (M. Wt. 157.5) adopted for this Process.

= 69.0% of Theory on the initial charge of 250.0 Kgs of real o-NCB (M. Wt. 157.5).

5. Raw Materials Usage Data:

(For 135.0 Kgs real o-Anisidine)

	Names of Raw Materials.	*M. Wt.*	*Quantity Real (Kgs)*	*Moles*	*One Mole Ratio*
i)	o-Nitro-Chloro-Benzene	157.5	250.0	1.5873	1.0000
ii)	Methanol, employed	32.0	625.0	–	–
	Methanol, used	32.0	200.0	6.2500	3.9375
	Methanol, recovered	32.0	750.0	–	–
iii)	Caustic Soda, Flakes, as is 98%	40.0	114.00	–	–
	Real 1100%	40.0	112.0	2.8000	1.7640
iv)	Hydrochloric Acid,				
	as is 30%	36.5	130.0	–	–
	Real 100%	36.5	39.0	1.0685	0.6731
v)	Iron Filings, as is	56.0	255.0	4.5536	2.8688
vi)	Formic Acid, real	46.0	8.5	0.1848	0.1164
viii)	Magnesium Oxide	40.0	11.0	0.2750	0.1732

8

Manufacture of 2:4-Dinitro-Aniline

1. Principle:

2:4-Dintro-Chlorobenzene is aminated with Ammonia Solution in an autoclave under pressure. The aminated batch mass is finally drowned into water, filterd and washed. The wet centrifuged product is evaluated for real content and used as such without drying.

2. Reactions:

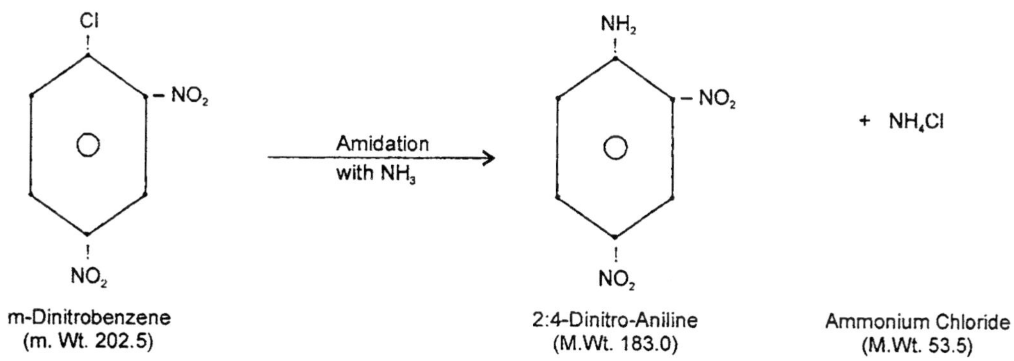

3. Process-Procedure:

3.1. Amidation

Tool A 3,000 L capacity Special MS Autoclave with MS Limpet Coil and Jacket and designed to give 150 p.s.i. Working Pressure. The Autoclave is equipped with an MS Gate-type Agitator driven by a 5 H. P. closed Electric motor connected with a Reduction-gear to give 60 r. p. m. speed to the stirrer. The Autoclave is further provided with a Manhole, a Temperature-recorder, a Pressure-gauge and Blow-out line leading to the Drowning Kettle.

Check tool

In the clean Autoclave,

Charge

1,735.0 L of as is 15%/w/v Ammonia solution corresponding to 260.0 Kgs of real Ammonia (M. Wt. 17.0).

Start agitator and charge in it through the open Manhole

640.0 Kgs of real, molten 2:4-Dinitro-Chlorobenzene (M. Wt. 202.5), previously melted to 70-75°C with steam in special jacketed Melting Pots.

Note : Generally instead of using Melting Pots, the MS drums containing solid, fused DNCB are placed in wooden cases open from above and a passage of direct high-pressure steam is projected around the external walls of the drums, until the solid material in the drums gets melted.

The charging of molten DNCB takes about 1-1/2 to 2 hours. During the hot molten DNCB charging, no temperature-control in the Autoclave is necessary.

At the end of this charging, check volume of the total charge in the Autoclave; it will be about

2,000.0 L

Close the Manhole and seal the Autoclave to assure no leaks.

Continue stirring.

Now blow a passage of steam in the limpet coil and cautitiously heat the charge to 70°C regulating the temperature rise at a rate of 7°/hour. The exothermic reaction sets in at about 70 to 85°C and hence, the sudden rise in temperature in this range has to be very carefully controlled by checking the steam passage.

Then heat the batch further to 115-120°C during about 4 to 5 hours, taking enough care for the self rise in temperature with the generation of exothermic heat.

At 115-120°C the temperature will be nearly stabilized and the pressure built up will be about 60-62 p. s i.

Stir the batch at this temperature and pressure for four hours.

By now, the amidation is positively complete.

Then stop steam on the Autoclave and cool and batch to 80-85°C, whereby the pressure in the Autoclave drops to 0 p. s. i.

The batch is now ready for drowning.

3.2. Drowning and Filtration

Tool · A 12,000 L capacity M. S. jacketed Kettle equipped with an M. S. spade-type Agitator driven by a 5 H. P. closed Electric motor connected with a Reduction-gear to give 50 r. p. m. speed to the Agitator. The Kettle is further equipped with a Manhole, a Thermometer a Direct steam line, a water-feed line, a Bottom-discharge line and a Blow-out line leading to the Product Nutsche Filter.

Check tool.

Before the batch in the Autoclave is ready for drowning, charge into this Kettle

5,000.0 L of water.

Start agitator and receive into it the batch mass held at 80-85°C in the Autoclave via the blow-out line.

Do not delay in transferring the batch from the autoclave, once the blowing is started.

To remove the adhered material in the empty Autoclave, give two washings to it, each washing corresponding to

1,000.0 L of 2% Ammonia solution that is stirred in the Autoclave holding the main batch mass.

Stir the batch in the Drowning Kettle and record the temperature.

Inject direct steam in the batch and heat to 80-85°C.

Stir for half hour at 80-85°C.

The batch is now ready for filtration.

Record the volume of the drowned, batch; it will be about

9,00.0 L at 80-85°C.

Blow the batch into the Nutsche Filter at 80-85°C and filter fast.

Suck the M. L.

The M. L. contains not more than 0.1% of the amine. Discard the M. L. to drains after giving it the relevant treatment as may be necessary.

Suck the nutsche cakes well.

Finally, give two washings to the Nutsche cakes, each time with

700.0 L of ordinary water.

Suck the washings and release to drains.

Press the cakes and suck well.

Dump the nutsche.

Finally, centrifuge the Nutsche cakes.

Weigh the centrifuged cakes and record.

Draw a well prepared uniform sample of the wet cake and determine its 2:4-Dinitro-Aniline content with a simple Nitrite titre.

Record yield.

The product is not dried and used in the form of wet cakes.

4. Yield:

Weight of centrifuged Wet Cakes,

= Approx. 650-675 Kgs

= 500.0 Kgs of real 2:4-Dinitro-Aniline (M. Wt. 183.0).

= 86.5% of Theory on the charge of real 2:4-Dinitro-Chlorobenzene (M. Wt. 202.5).

5. Raw Materials Usage Data:

	Names of Raw Materials.	*M. Wt.*	*Quantity Real (Kgs)*	*Moles*	*One Mole Ratio*
i)	2:4-Dintro-Chlorobenzene	202.5	640.0	3.1605	1.0000
ii)	Ammonia	17.0	260.0	15.2941	4.8391

Cleaning the Autoclave

At the end of the use, charge the empty Autoclave with about 2000/2200 L of water. Close the Autoclave and heat the water to 130-140°C. Stir for 15-20 minutes at 130-140°C and blow out the washing. Then fill the clean Autoclave with ordinary water, stir for half an hour and drain down the water.

9

Manufacture of 4-Amino-Acetanilide-3-Sulfonic Acid

1. Principle:

p-Nitro-Aniline-*o*-Sulfonic Acid is reduced with iron borings and Ferrous chloride pickle liquor in a neutral aqueous medium. The reduced product is sodiated and filtered free from iron oxide. The solution of Sodium salt of *p*-Phenylenediamine-*o*-Sulfonic Acid is acetylated with Acetic anhydride and the resulting objective product is isolated by acidification with HCl. The precipitated product is filtered and the wet cake evaluated for real content.

2. Reactions:

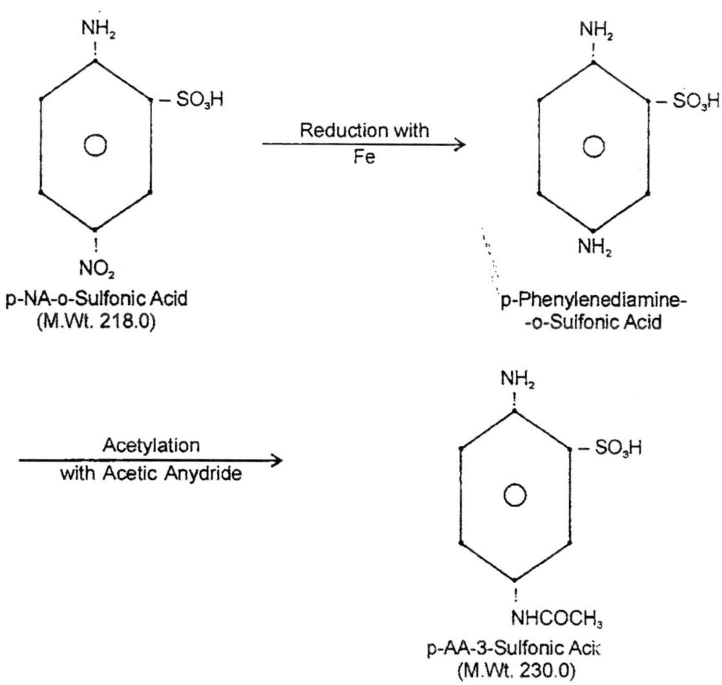

3. Process-Procedure:

3.1. Reduction

Tool A 6000 L capacity C. I. Kettle with bottom-section concrete-lined or an M. S. R. T. L. Open Vessel with a top Wooden Cover and a charging-door on it. The tool is equipped with a C. I., Propellar-type agitator driven by a 5 H. P., 3-Phase, closed Electric motor connected with a Reduction-gear to give 40 r. p. m. speed to the agitator. The tool is further provided with a Temperature-recorder, a direct steam line, a Vent line with an Exhaust fan to drive off ammonia and bottom-outlet line leading to the Filter Press.

Check tool.

In the clean Reducer, charge

3000 L of water.

Start agitator and charge a total quantity of

565 Kgs of Iron borings and

380 Kgs of Ferrous Chloride Pickle-liquor.

Note : In the subsequent batches, 195 Kgs of recovered Iron borings from the previous batch are charged and the balanced quantity is made up with fresh Iron borings. Also, the initial water-charge in the reductions of the following batches is made up using the end washings of the Iron-oxide sludge and this has to be made slightly acidic (to no test on B. Y. papers) by adding little HCl, if it is alkaline.

Heat the charge to 92-95°C open steam and stir at this temperature for 1 hour.

Then evenly charge over a period of 6 hours

567 Kgs of real p-Nitro-aniline-o-Sulfonic Acid (M. Wt. 218.0) in the form of Wet cakes, maintaining the bath temperature at 95-97°C by the heat of reaction.

At the end of the addition of the Wet cakes, stir the batch for 3 hours at 95-97°C by passing direct steam in the batch.

Draw a sample of the batch and check for the completion of reduction as follows:

Test : Dilute 100 ml of the batch sample to 200 ml volume with hot distilled water and heat to boil. Add to it 20% Soda Ash solution until alkaline to B. Y. papers. Filter the solution through a Buchner funnel. Measure 10 ml of the clear filtrate in a Reduction flask. Add to it 20 ml of 50% H_2SO_4 solution and 25 ml of a Standard N/5 Titanous trichloride solution.

Bubble CO_2 gas in the mixture, boil for 10 minutes and cool to room temperature. Add 5 ml of 1% Ammonium Thiocyanate indicator solution and back-titrate the excess $TiCl_3$ with a Standard N/5 Ferric Alum solution passing CO_2 gas throughout the operation. Note the burette reading and call it "X" ml.

Run a blank titration using all the above reagents and substituting the 10 ml of sample filtrate with 10 ml of distilled water and observing similar conditions.

Call this burette titre as "Y" ml.

If the titre "Y" exceeds the titre "X" by more than 0.2 ml, the reduction is still incomplete.

If the reduction of the batch is still partially incomplete, stir the batch further for 1 hour at 95-97°C and retest as above. During the testing period, the reduction should be allowed to go on. Total maximum reduction period right from charging the wet cakes = Approx. 12 hours.

The reduction is complete and the batch is ready for Sodiation.

3.2. Sodiation

Tool In the Reducer itself.

When the reduction is complete, shut off steam in the Reducer. Start the exhaust fan on the vent line. Carefully add Soda Ash powder to the batch avoiding excessive frothing to adjust the batch pH in the range of 9.2-9.5.

Test a batch sample for the absence of soluble iron.

Test : Spot a drop of the batch sample on a piece of filter paper. Cross-spot the clear run-out with a drop of 1% Potassium Ferricyanide solution. Appearance of no blue line at the junction of the two spots indicates the absence of soluble iron.

Sodiation requires about

145 Kgs of Soda Ash powder.

When the soluble iron is confirmed to be absent, the Sodiation is treated as complete.

The charge in the batch

8 Kgs of Sodium Hydrosulfite and stir the batch for ½ hour to drive off Ammonia gas through the exhaust Vent. Total time for Sodiation = Approx. 2 hours.

The batch is now ready for filtration.

3.3. Filtration

Filter the batch hot through the Filter Press.

Receive the batch filtrate into the Acetylation Vessel. Quickly cool the batch filtrate received in the Acetylation Vessel to 28-30 °C.

Rinse the empty Reducer with about

1200 L of hot water of 70-75 °C and pump it into the Press to wash the iron-oxide sludge. Collect the sludge washings along with the main batch filtrate in the Acetylation Vessel.

Then further wash the empty Reducer and the Press sludge with sufficient quantities of hot water until the total volume of the main batch filtrate and washings in the Acetylation Vessel is adjusted to

5600 L at 28-30 °C.

Finally, give a further wash to the iron oxide sludge with about

2800 L of hot water and reserve this washing in the Reducer for making up the initial water charge for the next batch. Remove the alkalinity if any of this washing with a little HCl to no test on B. Y. papers. Air-dry the Filter Press for 2 hours.

Discharge the iron-oxide sludge from the Press

Test a sludge sample for loss of Amino compound as follows:

Test : Transfer 100 gms of the sludge sample into a 1 L beaker. Add about 500 ml of water, stir and heat to boil. Filter through a Buchner funnel and wash the cakes with 100 ml of hot water.

Transfer the filtrate and washings in a 1L beaker cool to 20 °C and add 25 ml of Conc. HCl. Titrate with a Standard N/10 Sodium Nitrite solution to a 3-minute test on KI-Starch papers. The titre should not exceed 20 ml of N/10 $NaNO_2$ solution which corresponds to about 0.03% loss of the Amino compound on theory yield.

3.4. Mono-acetylation

Tool An 8000 L capacity M.S. R. L. Kettle or preferably a Wooden Vat with a top Wooden cover having a charging-door on it and equipped with a wooden, Gate-type agitator driven by a 5 H. P., 3-Phase, closed Electric motor connected with a Reduction-gear to give 30 r. p. m. speed to the agitator.

The Vat is further provided with a temperature-recorder and a bottom-outlet line leading to the Filter Press.

The clear batch filtrate along with the washing from the Sodiation Step is adjusted to

5600 L volume at 28-30 °C, quickly after receipt as said earlier.

Start agitator and neutralize the excess alkalinity of the batch by adding to it as required, about

90 Kgs of as is 30%w/w HCl corresponding to 27 Kgs of real HCl to not test on B. Y. papers.

This neutralisation of alkalinity also should be carried out immediately after receiving and cooling the batch liquor in this Vat, without making any delay.

Draw a sample of the batch solution and determine real p-Phenylendiamine-o-Sulfonic Acid content on M. Wt. 188.0 as described below.

Test : Measure 50 ml of sample and dilute it to 250 ml volume with distilled water in a Volumetric flasks.

Pipette 50 ml of this volumetric solution into a 500 ml beaker. Stir the solution add to it 5 gms of Oxalic acid crystals to make the solution acidic to Congo Red test papers. Cool to 05°C by adding washed ice. Then under stirring, titrate it with a Standard N/10 Sodium Nitrite solution to one-minute end point (blue test) on KI-Starch papers.

Repeat this titration by adding the N/10 $NaNO_2$ solution as quickly as possible to within 1 ml of the burette titre (obtained in the first titration) and finish off the rest of the titration to get the end-point.

Use the second titre reading for calculations.

Let the burette reading

is "A" ml of N/10 $NaNO_2$ solution.

Then,

the total, real *p*-Phenylenediamine-o-Sulfonic Acid content in the entire batch volume,

$= A \times 0.0188 \times 250/50 \times$ Batch Vol. in L./50

$= (A \times 0.00188 \times$ Batch vol. in L.) Kgs on M. Wt. 188.0

Note : From the above test, determine the charge of real Acetic Anhydride to the order of 1.06 Kg Mole of Acetic Anhydride per 1 Kg. Mole of *p*-Phenylenediamine-o-Sulfonic Acid, i.e. 108 Kgs of real Acetic Anhydride for 188 Kgs of *p*-Phenylenediamine-o-Sulfonic Acid. Generally, the conversion to be 100% of Theory yield and the Acetic Anhydride charge is determined on his basis.

Measure the calculated quantity of Acetic Anhydride amounting to

282 Kgs of real Acetic Anhydride (M. Wt. 102.0) in a Measure Vessel and charge it into the batch under the surface of the agitating batch liquor during about 1 hour maintaining the batch temperature in the range of 29-

32°C, throughout the addition. Stir the batch for 1 hour at 30-32°C at the end of addition of Acetic Anhydride.

Draw a batch sample at this stage and test for the completion of Acetylation as follows.

Test : Measure 25 ml of the batch sample in a 100 ml beaker. Add to it 5 ml of re-distilled Cellosolve and 15 ml of 1N HCl solution. Sir well with a glass rod and filter. Dilute the filtrate to 100 ml volume with distilled water.

Pipette 10 ml of this dilute solution in a 100 ml beaker. Add to it 1 ml of 1 N HCl solution, cool to 10-12°C and add 1 ml of N/2 Sodium Nitrite solution. Stir with a glass rod and allow to stand for 5 minutes. Add this solution to a mixture of 5 ml of 10% Soda Ash solution, 5 ml of M/20 R-Salt solution and 30 ml of distilled water. Stir the mixture well with a glass rod. Develop a Chromatogram using this solution on a special Chromatography paper held between two glass plates using distilled water as a developer.

There should be no blue inner ring in the Chromatogram.

This indicates completion of acetylation.

If the acetylation is still partially incomplete, add Acetic Anhydride in 5-lb lots and stir the batch at 30-32°C until the above test is satisfied.

The acetylation is complete.

Measure volume of the batch; it will be about

6000 L at 30-32°C.

The batch is now ready for isolation.

Notes : (1) The batch under acetylation is partly out of solution after about half the quantity of Acetic anhydride is charged. On addition of the total charge of Acetic anhydride, the batch goes into solution.

(2) Under acetylated material gives a high unacetylated diamine content in the finished product, whereas over-acetylation tends to give lower yields.

3.5. Isolation and Filtraion:

Tool Same as Acetylation Vessel.

While stirring, add during about 15-20 minutes

550 Kgs of as is 30%w/w HCl corresponding to 165 Kgs of real HCl.

Stir for 1½ hours without temperature control.

The product precipitates out.

Draw a sample of the batch and determine the liquor-loss content as follows.

Test : Filter the sample through Buchner funnel.

Pipette 25 ml of the clear filtrate in a 500 ml beaker containing 200 ml of distilled water of 5-8°C.

Stir and add 10 ml of 30%w/w HCL.

Titrate the solution with a Standard N/10 Sodium Nitrite solution to 1-minute end point on KI-Starch papers.

Note the burette titre.

Then,

(ml. N/10 $NaNO_2$ titre × 0.092)

= % Liquor-loss by volume.

Calculate total liquor-loss in the batch volume as percentage of Theory. The liquor-loss is usually below 8% of Theory.

If the liquor-loss is above 12% of Theory, stir the batch further for 2 hours at 30°C and retest as above. Total time for Isolation = about 4-1/2 hours.

The batch is now ready for filtration.

Filter the batch through Filter Press at Room-temperature. Suck the cakes well and discard the M. L. to drains.

Air the Press for 2 hours.

Dump the wet product from the Press.

Weigh the wet cakes and record.

Determine real content in a representative sample of the wet product and calculate yield.

4. Yield:

Weight of Wet Cakes,

= Approx. 1000-1100 Kgs.

= 500 Kgs of Real *p*-Amino-Acetanilide-3-Sulfonic Acid (M. Wt. 230.0).

= 83.6% of Theory on *p*-NA-o-Sulfonic Acid charge.

5. Raw Materials Usage Data:

	Names of Raw Materials.	M. Wt.	Quantity Real (Kgs)	Moles	One Mole Ratio
i)	*p*-Nitro-Aniline-*o*-Sulfonic Acid (About 70/75% wet cakes)	218.0	567.0	2.6009	1.0000
ii)	Iron Borings, screened	–	565.0	–	–
iii)	Ferrious Chloride, Pickle liquor	–	380.0	–	–
iv)	Soda Ash, powder	106.0	145.0	1.379	0.5259
v)	Sodium Hydrosulfite	–	8.0	–	–
vi)	Hydrochloric Acid, real	36.5	27.0	0.7397	0.2844
vii)	Acetic Anhydride	102.0	282.0	2.7647	1.0629
viii)	Hydrochloric Acid, real	36.5	165.0	4.5205	1.7380

10

Manufacture of 2:5-Dichloro-Nitrobenzene

1. Principle:

p-Dichlorobenzene is nitrated with Mixed Acid in Sulfuric acid medium. The nitrated mass is drowned in water and washed several times with hot water to remove most of the excess acid. Finally, the remnant excess acidity in the product is neutralized with Soda Ash and the product filtered, washed and dried.

2. Reactions:

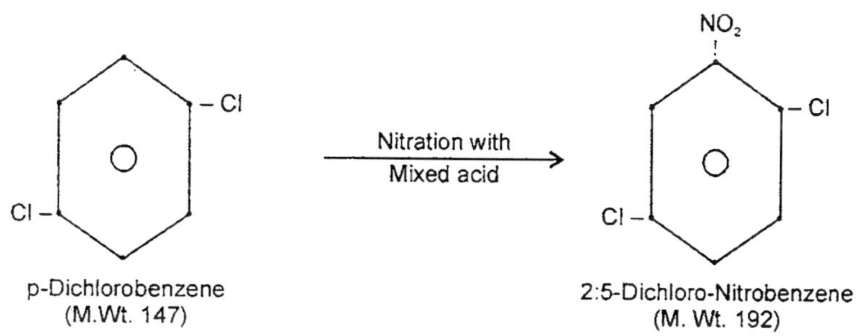

3. Process-Procedure

3.1. Preparation of Mixed Acid:

Tool A 5000 L capacity, SS-lined M. S. jacketed Vessel equipped with an Anchor-type agitator driven by a 3 H. P., 3-Phase, closed Electric motor connected with a Reduction-gear to give 60 r. p. m. speed to the agitator. The Vessel is further provided with a Manhole, a Vent-line, a Temperature-recorder and a Blow-out line.

Check tool.

Charge in the clean, dry Kettle

2650	Kgs	of as is 98%w/w Sulfuric Acid corresponding to
2597	Kgs	of real Sulfuric Acid.
		Start agitator.
		Circulate chilled water in the Kettle jacket and cool the acid to 10°C.
		Then add gradually, holding the temperature below 30°C
2547	Kgs	of as is 93%w/w Nitric Acid corresponding to
1700	L	volume and containing
2369	Kgs	of real Nitric Acid (M. Wt. 63.0).
		Stir the mixture at 30-32°C for 1 hour.
		Measure total weight and volume of the Acid.
5197	Kgs	of total weight and
3140	L	total volume of the Mixed Acid at 30°C.
		This Mixed Acid contains
45.5%/w/w		as real Nitric Acid (M. Wt. 63.0).

3.2. Nitration

Tool : A 1500 L capacity, C. I. jacketed Kettle equipped with an Anchor-type Agitator driven by a 3 H. P., 3-Phase, closed Electric motor connected with a Reduction-gear to give 40 r. p. m. speed to the Agitator.

The Kettle is provided with a Manhole, a Vent and a Blow-out line leading to the Drowning Vessel.

Check tool.

In the clean, dry Nitration Kettle, charge

310	Kgs	of as is 98%w/w Sulfuric Acid corresponding to
304	Kgs	of real Sulfuric Acid.
		Start agitator.
		Circulate chilled water in the Kettle jacket and cool the acid to 10°C.
		Then charge gradually
407	Kgs	of real *o*-Dichlorobenzene (M. Wt. 147.0) as broken lumps allowing the temperature not to exceed 20°C.

Note : By this time, keep ready 780 Kgs of as is 45.5% Mixed Acid containing 355 Kgs of real Nitric Acid, weighed in an SS Measure Vessel.

Stop circulation of cooling water in the Jacket. Stir for 10 minutes and gradually add some mixed acid from the above weighed quantity to rise the temperature of the charge to 30°C.

Then charge a IInd lot of

407 Kgs of real *p*-Dichlorobenzene (M. Wt. 147.0) as broken lumps allowing the temperature not to exceed 30°C by adjusting a controlled flow of chilled water in the Kettle jacket.

During this charging, the slurry tends to become thick and viscous.

Add some further quantity of Mixed acid from the measured charge to thin the slurry and keep it in a stirrable form.

Stop the flow of chilled water in the jacket. Add a further quantity of Mixed acid from the measured charge to raise the batch temperature to 50°C.

Finally, add the balance quantity of the Mixed acid holding the batch temperature between 50-55°C by controlling the flow of chilled water in the jacket. Stop circulation of chilled water in the jacket. Stop circulation of chilled water in the jacket at the end of addition of Mixed acid.

Then very carefully heat the batch to 68-70°C by passing a controlled passage of steam in the Kettle jacket.

Stir for 2 hours at 69-71°C.

Test for completion of Nitration as follows:

Test : Draw a sample of nitrated mass.

Drown about 100 gms of the sample in 1500 ml of ice-cold water placed under stirring in a 2-litre beaker. Add more washed ice as required to keep the temperature between 20-30°C. Stir for 10 minutes. Filter the precipitated product through a Buchner funnel and wash the cake with plenty of water to no blue test on Congo Red papers.

Suck the product well of drain down water.

Then quickly flush on the cake about 15-20 ml of Acetone and suck down. Press the cake well. Remove the cake and further dry the product by pressing between folds of filter paper. Determine Setting-Point of this product. This product must show a Setting-Point of 52.5-53°C.

If the above test is satisfied, the Nitration is treated as complete. If the Setting-Point test is not satisfactory, continue Nitration further for 1 hour at 69-71°C and retest as above.

3.3. Drowning:

Tool A 7500 L capacity, M. S. R. L. T. L . Kettle equipped with a Gate-type Agitator driven by a 5 H. P., 3-Phase, closed Electric motor connected with a Reduction-gear to give 40 r. p. m. speed to the Agitator.

The Kettle is further equipped with a Manhole, a Pressure-Cooling coil, a direct Steam line and a Bottom-outlet line Leading to the Pressure-Nutsche.

		Check tool.
		In the clean Drowning Vessel,
		Charge
4000	L	of water.
		Start agitator.
		Then receive into this the nitrated batch from the above step gradually during about 2 hours. Allowing the drowning temperature not to exceed 55 °C, by circulating chilled water through the Cooling-coil as necessary.
		Measure volume and adjust to
5500	L	at 58-60 by adding water and passing direct steam as is necessary.
		Stir for 10-15 minutes at 58-60 °C.
		Stop stirrer and allow the batch to settle for half an hour.
		The batch separates into two layers.
		Draw off the upper aqueous layer and discard to drains. Start agitator.
		Add water to the oily bottom layer and adjust volume to
2700	L	a 58-60 °C by simultaneously passing direct steam in the charge.
		Stir for 15-20 minutes at 58-60 °C.
		Stop stirrer and allow the batch to settle for half an hour.
		Again draw off the upper aqueous layer and discard to drains.
		This way, give one more washing to the lower oily layer of product making
2700	L	volume with water and steam at 58-60 °C.
		Finally, by adding water and simultaneously passing steam, make up the batch volume to
3000	L	at 58-60 °C.
		While stirring, add to the batch Soda Ash powder as required to obtain a faint-orange test on B. Y. papers.
		Stir for half an hour at 58-60 °C
		And confirm faint-orange test on B. Y. papers. Then while stirring, add requisite quantities of ice to cool the batch to 25 °C during about 2 hours.
		Stir for 1 hour at 23-25 °C.
		The product solidifies out in the form of fine crystals.
		The batch is now ready for filtration.
		Blow the batch to Pressure-Nutsche.

Suck M. L. well.

Wash the Nutsche cakes thrice, each time with

400 L of water. Discard the washings.

Suck the washing and press the Nutsche cakes.

Dump the Nutsche.

Centrifuge the Nutsche cakes to nearly dry state.

Weigh the centrifuged product and record.

Dry the product in Air-dryers strictly at

40-45 °C.

Weigh the dry product and record.

Determine Setting-Point of the dry product; it must be in the range of 52.5-54.0 °C.

4. Yield:

Weight of centrifuged product,

= Approx. 1225-1250 Kgs.

= Approx. 1020-1030 Kgs of dry product having

97.0-98.% real

2:5-Dichloro-Nitrobenzene

content (M. Wt. 192.0).

= 1000 Kgs of real 2:5-Dichloro-Nitrobenzene (M. Wt. 192.0)

= 94% of Theory on real *p*-Dichlorobenzene charge.

5. Raw Materials Usage Data:

	Names of Raw Materials	*M. Wt.*	*Quantity Real (Kgs)*	*Moles*	*One Mole Ratio*
1.	*p*-Dichlorobenzene	147.0	814.0	5.5380	1.0000
2.	Sulfuric Acid, Real	98.0	304.0	3.1020	0.5603
3.	Nitric Acid real as Mixed Acid	63.0	355.0	5.6350	1.0170

11

Manufacture of 4-Chloro-Nitrobenzene-3-Sulfonic Acid

1. Principle:

p-Chloro-Nitrobenzene is sulfonated with Oleum and the product is isolated by drowning into a solution of Glauber Salt. The isolated product is filtered, washed with brine and the wet cakes evaluated for real content. The wet product is not dried but used as such for the conversion to *p*-Nitro-Aniline-*o*-Sulfonic Acid.

2. Reactions:

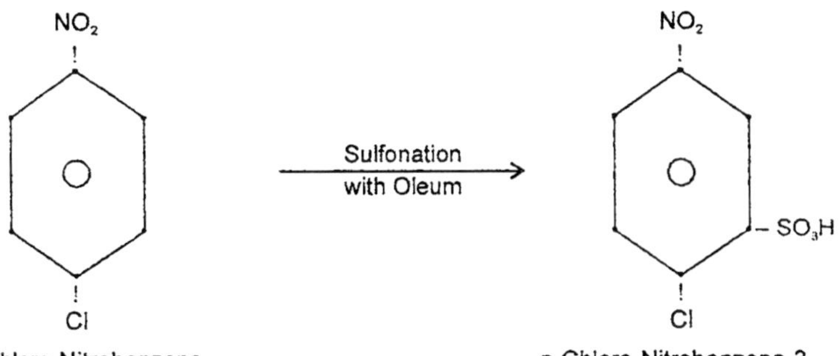

p-Chloro-Nitrobenzene (M.Wt. 157.5) → p-Chloro-Nitrobenzene-3-Sulfonic Acid (M.Wt. 237.5)

3. Process-Procedure:

3.1. Sulfonation

Tool A 1000 L capacity C. I. Kettle with an M. S. Jacket equipped with an Anchor-type agitator driven by a 3 H. P., 3-Phase, closed Electric motor connected with a Reduction-gear to give 40 r. p. m. speed to the agitator.

The Kettle is further equipped with a manhole, a temperature-recorder, a vent line connected with the fume-absorption system and a blow-out line leading to the Isolation Vessel.

Check tool.

In the clean, dry Sulfonation Kettle, charge

120 Kgs of 25-27% Oleum from Measure Vessel.

Start agitator and vent the Kettle to the fume-absorption system.

Heat the Oleum to 60°C with jacket steam.

Then charge gradually during about half-hour

350 Kgs of real, previously molten *p*-Chloro-Nitrobenzene (M. Wt. 157.5).

The temperature tends to rise to 80-82°C.

If the temperature-rise tends to overshoot 82°C,

Circulate cold water in the kettle jacket to control it at or below 82°C.

Then charge through the orifice plate

325 Kgs of as is 65% Oleum during about 2 to 2-1/2 hours, maintaining the temperature between 80-85°C by circulating chilled water through the jacket. At the end of Oleum addition, heat the batch with jacket steam to 115°C, uniformly during 2 hours as shown in the following pattern.

i)	From 85 to 95°C	Turn off steam and allow the temperature to rise by itself to 95°C.
ii)	From 95 to 102°C	In the same way as above if the self-rise is maintained.
iii)	From 102 to 114°C	Regulate the temperature-rise by passing controlled passage of steam in the jacket.

Then stir the sulfonation at 114-116°C for 2 hours. Draw a sample of the batch and test for completion of sulfonation as follows:

Test : Take about 25 gms of crushed ice in a clean 200 ml beaker. Drop in it about 10 ml of the sulfonation sample while stirring with a rod. Then add to it as required 25% Caustic Soda solution until a distinct red test on Phenolphthalein papers is obtained. Further dilute the solution with water, if necessary, to 100 ml volume and heat to 90°C.

The solution must be clear and should show no oily globules of unsulfonated and molten *p*-Chloro-Nitrobenzene. Unsulfonated *p*-Chloro-Nitrobenzene also emits peculiar aromatic odour from the solution.

If the sulfonation is still incomplete at the end of the 2 hours reaction at 115°C, stir the batch further 1 hour at 114-116°C and retest for the completion of sulfonation as above.

When sulfonation is complete, stop heating and cool the batch to 80°C by circulating water through the jacket. The batch is how ready for isolation.

Measure volume of the batch; it will be about

490 L at 80°C.

3.2. Isolation

Tool A 4000 L capacity M. S. L. L., jacketed open Vessel with a paddle-type lead-lined agitator driven by a 5 H. P., 3-Phase, closed Electric motor connected with a Reduction-gear to give 40 r. p. m. speed to the agitator. The Vessel is provided with a top Wooden cover having a charging-door on it. The Vessel is further equipped with a direct steam line, a temperature-recorder and a bottom-outlet line leading to the Filter Press.

Check tool.

In the clean Isolation Vessel, Charge

1950 L of treated water.

Start agitator and steam to 40°C with direct steam.

Then charge in it

390 Kgs of real Glauber Salt.

Stir at 40-45°C to complete solution. Add

5 Kgs of Dispersol OG or Calsolene oil HS.

Stir to solution.

Now receive into this Isolation charge the sulfonated batch from the above step, gradually during about half an hour, allowing the drowning temperature to rise to 75-80°C.

Record temperature.

Heat the batch to 85°C with direct steam.

Measure volume; it will be about

2700 L at 85°C.

Then stop heating an stir the batch to cool to 60°C. This takes about 8-9 hours.

Then further cool the batch to 30°C during about 2 hours by circulating chilled water through the Vessel jacket.

Charge

110 Kgs of real Glauber Salt in the batch over a period of about 15-20 minutes.

		Stir at 28-30°C for 1 hour.
		Finally, cool the batch further to 20-22 °C by circulating chilled water through the jacket.
		Stir the batch for 1 hour more at 20-22°C.
		Draw a sample, filter it and determine the loss of product in the M. L.
		The isolation is positively complete and the liquor-loss does not exceed 4.5% of Theory.
		The batch is now ready for filtration.
		Blow the isolated batch into the Nutsche filter having 7.4 Sq. meter basket area.
		Suck the M. L.
		Rinse the empty Isolation Vessel with about
100	L	of as is 30%w/vol Brine. Stir and transfer this washing over the wet cake in the Nutsche. Suck the Washing. Further wash the nutsche cake twice, each time with
200	L	of as is 30%w/vol Brine allowing the washing to soak into the wet cakes for about 15 minutes before sucking.
		Suck the washings.
		Discard the M. L. and washings to drains.
		Press the nutsche cakes well under sucking.
		Dump the wet cakes from the Nutsche.
		Centrfuge the nutsche cakes.
		Dump the centrifuged cakes, weigh and record.
		Draw a representative sample of the centrifuged
		Product and determine the real content in it.
		Calculate yield.

4. Yield:

Weight of centrifuge wet product,

= Approx. 750-800 kgs.

= 500 Kgs of real *p*-Chloro-Nitrobenzene-3-Sulfonic Acid (M. Wt. 237.5).

= 94.74% of Theory on p-Chloro-Nitrobenzene charge.

5. Raw Materials Usage Data:

	Names of Raw Materials	*M. Wt.*	*Quantity Real (Kgs)*	*Moles*	*One Mole Ratio*
1.	*p*-chloro-Nitrobenzene	157.5	350.0	2.2264	1.0000
2.	Oleum, 25/27%	–	120.0	–	–
		80.0SO_3:	96.0	1.2000	0.5389
3.	Oleum, 65%	–	325.0	–	–
		80.0 SO_3:	260.0	3.2500	1.4597
4.	Glauber Salt	142.0	500.0	–	–
5.	Calsolene Oil HS	–	5.0	–	–
6.	Salt in Brine	–	150.0	–	–

12

Manufacture of m-Ureidoaniline-Hydrochloride

1. PrincipleL

m-Phenylene Diamine (MPD) is dissolved with water and Hydrochloric Acid and the solution is reacted with Sodium Cyanate powder under controlled temperature and pH conditions. The precipitated *m*-Ureidoaniline-Hydrochloride is finally filtered and the wet product washed with neutral brine. The wet product is evaluated for real content and is used or sold as such, in wet form.

2. Reaction:

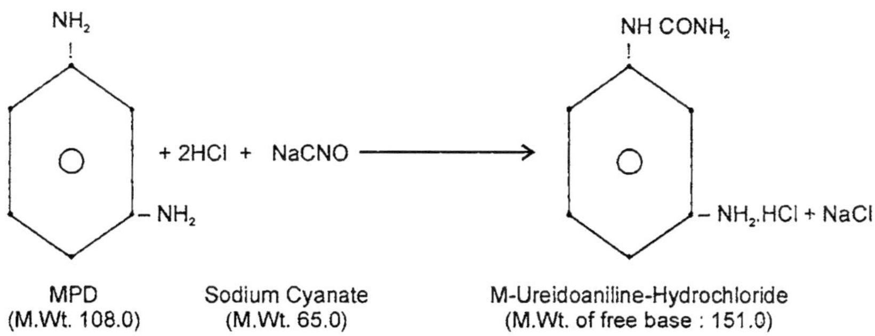

3. Process-Procedure:

3.1. Reaction with Sodium Cyanate:

Tool : A 3000 L capacity M. S. Glass-lined jacketted Kettle equipped with an anchor-type, Enamel-lined M. S. Agitator driven by a 5 H. P., 3-phase, closed Electric motor connected with a Reduction-gear to give 40 r. p. m. speed to the agitator. The Kettle is further equipped with a Temperature-Recorder cum Control, a Pressure-gauge, a vertical Vent line connected with a powerful Exhaust fan to drive off flue gases above the roof of the plant building, a manhole, an acid-charging line and a blow-out line leading to the Centrifuge or Vacuum Filters.

		Check tool.
		In the clean Kettle,
		Charge
725.0	L	of treated water.
		Start agitator.
		Then charge
795.0	Kgs	of as is 30% freshly analysed HCl, corresponding to
238.5	Kgs	of Real HCl.
		Open the vent of the Vessel and start the exhaust fan on the Vent line.
		Then charge
410.0	Kgs	of Real *m*-Phenylene Diamine (M. Wt. 108.0)
		having a Setting Point of 60-62°C. See Process Note (ii). During MPD charging, the temperature tends to rise upto 40-43°C.
		Stir the charge at 45-50°C (by external heating) until MPD dissolves completely. Frequently draw samples of the solution to observe that the solubilisation is complete.
		Cool the clear MPD solution to 20°C (Not below 19°C) by circulating chilled water in the Kettle jacket. As soon as the temperature comes down to 20°C, start charging Sodium Cyanate as follows.
		Before charging the Cyanate, read Process Note (ii).
		Prolonged delay in Cyanate charging may cause the MPD-Hydrochloride to fall out at 20°C.
		Charge
7.5	Kgs	of Pine Oil and follow with the charging of
180.0	Kgs	of Real Sodium Cyanate powder (M. Wt. 65.0), completing the addition during 4-4-1/2 hours, strictly at 19-21°C. During the Cyanate charging, the pH of the batch must be strictly maintained in the range of 1.7-2.3. If more, correct it by adding extra quantity of 30% HCl as required. If this addition is made, record the quantity of acid added. The reaction mass must show a blue colour on Congo Red papers at all times.
		Stir the batch for 45 minutes at 19-21°C, holding the pH in the range of 1.7-2.3.
		Then at an interval of every 1-1/2 hours add further quantities of Sodium Cyanate and Hydrochloric Acid in six lots, in the sequence described below.

Pattern of Cyanate and HCL additions:

Ist Lot..... 18 Kgs of 30% HCl and
16 Kgs of Real Sodium Cyanate.

2nd Lot ... 18 Kgs of 30% HCl and
16 Kgs of Real Sodium Cyanate.

3rd Lot.... 18 Kgs of 30% HCl and
16 Kgs of Real Sodium Cyanate.

4th Lot.... 18 Kgs of 30% HCl and
17 Kgs of Real Sodium Cyanate.

5th Lot.... 18 Kgs of 30% HCl and
17 Kgs of Real Sodium Cyanate.

6th Lot.... 18 Kgs o 30% HCl and
17 Kgs of Real Sodium Cyanate.

During the Addition of these six lots of HCl and Sodium Cyanate, the batch temperature should be maintained between 19-21°C and the pH in the range of 1.7-2.3 After the Cyanate addition is over, stir the batch further for 1-1/2 hours, strictly holding the temperature and pH in the said ranges. The reaction is almost complete or near completion.

m-Ureido-Aniline-Hydrochloride falls out in a fine crystalline form.

Measure batch volume and record.

Determine the Liquor-loss in the batch samples as described below and correct the batch accordingly until the liquor-loss comes within the ideal limit of 1.5-2.0% of theory on MPD charge (M. Wt. 108.0).

Determination of liquor loss:

Draw a sample of the batch slurry.

Measure exactly 600 ml of the well stirred sample and filter it through a dry buchner funnel and flask, Press the cake well to no drip. Do not wash the cake with water. Measure volume of the clear filtrate and call it "A" ml.

Pipette with bulb (Never with mouth, because it is hazardous) 100 ml aliquot of the clear filtrate into a 2-L beaker. Add to it about 800 ml of cold water of 10°C. Stir mechanically and add 5 gms of C. P. Sodium Bromide followed with 35 ml of Conc. C. P. HCL.

Titrate this solution at 10-12°C with a Standard N/1 Sodium Nitrite solution to a 3-minute end point on KI-Starch papers. Note the Burette reading and call it "B" ml.

Let "C" be the batch volume in Litres.

Then,

Percentage Liquor-loss,

$$= \frac{B \times 0.0755 \times A \times C}{600 \times 573.24}$$

= % Liquor-loss of theory of MPD charge.

If the liquor-loss is greater than 2.%, then add to the batch 1 kg of extra Sodium Cyanate for every 0.1% extra liquor-loss over 2.0%, followed with 1 kg of as is 30% HCl. Stir the batch at 19-21°C for 20 minutes holding pH between 1.7-2.3. Draw a batch sample and again determine the liquor-loss as above. The correction of the batch this way may be made until the liquor-loss falls in the range of 1.5-2%. When the liquor-loss falls in the range of 1.5-2.%. when the liquor-loss is adjusted in the said range, the batch is ready for filtration.

Caution:

Before blowing the batch to Vacuum Filters, run the exhaust fan on the Vent at full speed for at least 15 minutes to drive off any HCN gas collected in the upper empty space of the Vessel. Presence of more than 6 p.p.m. of HCN gas in the Vessel atmosphere is highly toxic and poisonous and therefore, it should be completely evacuated. Completely avoid breathing the vapours that come out while pumping the charge into the Vacuum Filters. Instant deaths of operating personnel have been recorded in the past as a result of this inhalation.

3.2. Filtration

Blow the batch to Vacuum Filters laid with special Terylene Filter Cloth. Suck the filter cakes free of M. L.

Wash the Filter Cakes thrice, each time with

450.0 L of neutral 20% w/vol Brine (the neutral brine should impart no tests on Congo Red and B. Y. test papers. Addition of a little HCl to the brine removes red test on B. Y. papers).

Between each brine wash, suck down the cake from the previous wash.

Finally, suck the cakes as dry as possible.

Dump the wet cakes from the Vacuum Filters.

Centrifuge the Wet Cakes.

Dump the Centrifuge, weigh the wet product and record.

Prepare a representative sample by mixing several samples drawn from different locations of the wet cakes mass. Analyse this sample for real content and calculated yield of the batch from the total weight of the wet product. Fill the standardized wet product in suitable containers and remove to storage.

Note : Instead of Vacuum Filters (Nutsche). The use of Centrifuge (M. S. R. L.) for direct filtration of the batch is very convenient and safe and hence recommended.

4. Yield:

Weight of Centrifuged wet cakes of the Finished Product,

= Approx. 1000-1100 Kgs at about 45-50% strength.

= 500 Kgs of Real *m*-Ureidoaniline (M. Wt. 151.0).

The product obtained is as *m*-Ureidoaniline-Hydrochloride.

= 87.22% of Theory on real MPD charge.

5. Raw Materials Usage Data:

	Name of Raw Material	*M. Wt.*	*Real (Kgs)*	*Moles.*	*One Mole Ratio*
1.	*m*-Phenylene Diamine (MPD)	108.00	410.00	3.7963	1.0000
2.	HCl, Real	36.50	238.50	6.5342	1.7212
3.	Pine Oil	–	7.50	–	–
4.	Sodium Cyanate	65.00	279.00	4.2923	1.1306
5.	HCl, Real	36.50	32.40	0.8877	0.2338
6.	Salt in 20% Brine (Washing)	–	270.00	–	–

6. Process Notes:

(i) Do not use more HCl then the quantity specified in preparing MPD solution for the following reasons:

(a) Sodium Cyanate reacts preferentially with excess of HCl causing immense frothing which results in the loss of Cyanate.

(b) Excess of HCl causes precipitation of MPD-Hydrochloride which is not desirable.

(ii) Sodium Cyanate is very likely to contain Sodium Cyanide as an impurity. The Cyanate under use should not contain more than 0.002% Sodium Cyanide. With the interaction of HCl, the cyanide liberates Hydrocyanic Acid gas which is extremely poisonous and causes instant death.

So, while charging the Sodium Cyanate, a powerful exhaust should be kept to vent out the HCN gas. Strict precautions should be taken at the Cyanate charging stage to avoid hazards. Instant deaths of operating personnel have been recorded in the past.

(iii) The MPD-Cyanate reaction is slightly exothermic. If the reaction temperature through mistake shoots up to about 26-28°C, serious frothing occurs causing the decomposition of Sodium Cyanate to CO_2 gas and Amonium Chloride.

(iv) The insoluble by-product formed during the reaction is known to be *m*-Diureidobenzene and its formation is attributed to further interaction with Sodium Cyanate as well as to the shooting up of reaction pH above the specified limit.

13

Manufacture of o-Tolidine Base

1. Principle:

o-Nitro-Toluene is reduced with Zinc and Caustic Soda in an Ethanol-Solvent Naphtha medium. The hydrazo-toluene extract in solvent naphtha is dropped in Sulfuric acid solution and converted to *o*-Tolidine Sulfate. The precipitate *o*-Tolidine Sulfate is separated by filtration and charged in water from which free o-Tolidine Base is obtained by treatment with Soda ash solution.

2. Reactions:

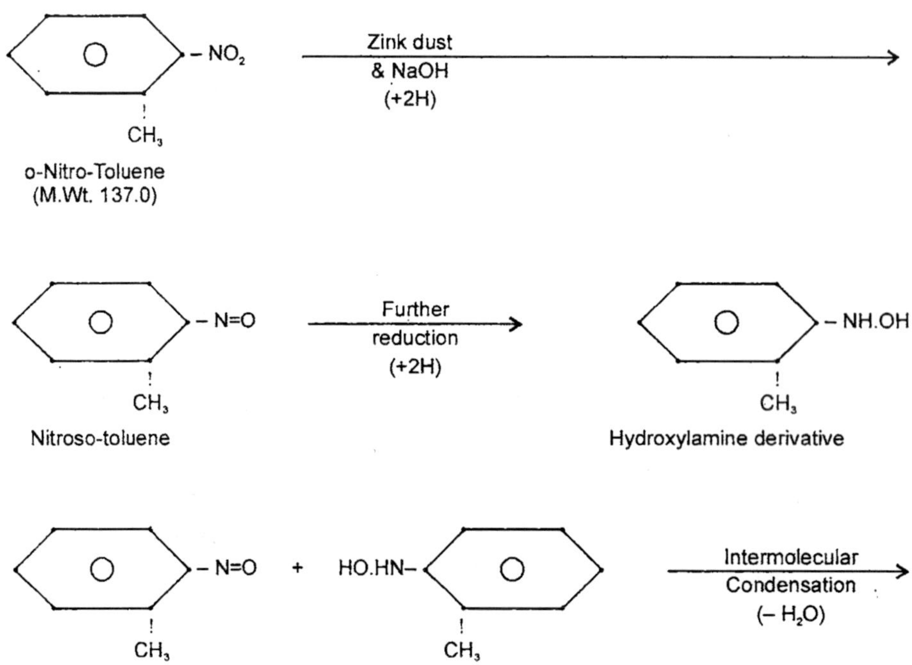

(Contd.)

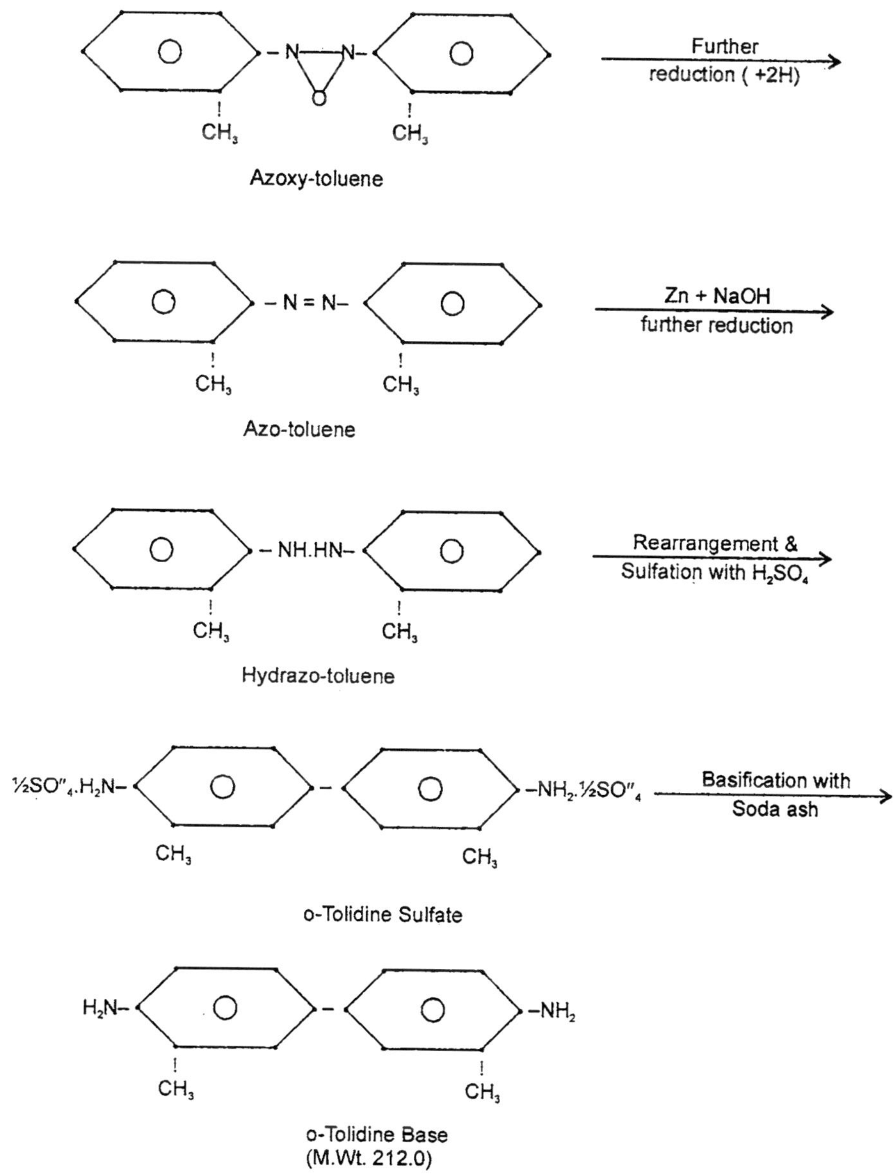

3. Process-Procedure:

3.1. Reduction to Azo-Toluene

Tool A 1500 L capacity SS Kettle with an M. S. jacket and equipped with an Anchor-type SS Agitator driven by a 2 H. P., 3-Phase, closed Electric motor connected with a Reduction-gear to give 40/30 r. p m. speeds to the agitator.

The Kettle is further provided with two diametrically opposite SS baffles, a vertical Shell and tube type Reflux Condensor, a horizontal Distillation Condensor, a temperature-recorder, a pressure-gauge, two Observation glasses, a manhole, a charging line and a bottom-outlet line

		In the clean Kettle,
		Charge
120	L	of Denatured Alcohol and
50	L	of Solvent Naphtha.
		Start agitator and charge
280	Kgs	as is 90% Zinc dust.
		Heat the charge to 35-40°C by circulating heated oil through the coil. Then charge under fast agitation
4	Kgs	of as is 50%w/w/ Caustic Soda lye corresponding to
2	Kgs	of real Caustic Soda, followed with
70	Kgs	of real o-Nitro-Toluene (M. Wt. 137.0).
		Circulate cold water in the jacket of the Reflux condenser.
		Heat the charge further to 80-82°C When reflux begins.
		Reduction starts at this stage.
		Continue the reduction further by making three alternate additions of 49% Caustic Soda and o-Nitro-Toluene, each addition lot consisting of
5	Kgs	of as is 50% Caustic Soda lye (=2.5 Kgs real NaOH) and
30	Kgs	of real o-Nitro-Toluene (M. Wt. 137.0).
		(Density of o-Nitro-Toluene is 1.161 of Denatured alcohol and Solvent Naphtha is 0.8).
		Till the end of the last addition, the reduction reaction should never die down.
		When the reduction reaction dies down at the end of all the above additions, the reduction uptil Azo stage is treated to be complete. Measure volume of the batch at this stage; it will be about
500	L	at 80-82°C.
		Test for the completion of reduction as follows:
		Test : Draw a small batch sample.
		Add about 3 to 4 ml of the sample to 50 ml of clear, colourles Benzene in 50 ml Nessler tube and mix up. If no red or orange colour is imparted to Benzene, the reduction is considered to be complete.

3.2. Reduction to Hydrazo-Toluene

When the reduction to Azo-Toluene is complete, stop reflux and open the valve on the Distillation line. Circulate cold water through the jacket of the distillation Condensor.

Distil off Alcohol at 80-90°C.

Then charge in the Kettle,

300 L of Solvent Naphtha.

Continue the distillation further until the charge attains a temperature of 95°C.

Then stop distillation and start only the reflux by circulating cold water through the jacket of the vertical Reflux Condensor.

Measure the volume of the Alcohol distillate; it will be about 150-175 litres.

Maintaining the Internal temperature at 95°C, reduce the Azo-Toluene to Hydrazo-Toluene by adding small lots out of a total quantity of

65 Kgs of as is 90% Zinc dust, each lot of Zinc dust being followed with the addition of a small lot of water, thus amounting to a total quantity of

14 L of water.

To control the temperature-rise as a result of the exothermic reaction, Solvent Naphtha may be intermittently added in lots to make up a total addition of

150 L of Solvent Naphtha.

The reduction reaction is fairly vigorous.

At the end of these additions, maintain the reaction at 95°C until the test for completion of reduction cited above is satisfied. When the reduction is complete as tested above, charge

300 L of Solvent Naphtha and stir the batch at 95°C under reflux so as to effect maximum extraction of the hydrazo compound in the solvent.

The while stirring at fast speed, add slowly to the batch a total quantity of

94 L of water so as to fix up the Zinc Oxide to hydroxide. Stir for 10-15 minutes at 90-95°C.

Stop stirring and measure volume of the batch; it will be about

1300 L at 90-95°C.

Allow the batch to settle for half an hour at 95°C. Then siphon of the clear upper layer of Naphtha solution of Hydrazo-Toluene straight into the Conversion Kettle holding the H_2SO_4 solution charge described as under.

Add to the Kettle

300 L of fresh Solvent Naphtha to further extract the Hydrazo compound remaining in the zinc hydroxide sludge.

Start the agitator very carefully, first with hand, then at low speed and finally at fast speed because the settled zinc hydroxide sludge is likely to bring load on the agitator which might damage it.

Heat the batch to 95°C and stir at 95°C for 15 minutes.

Stop stirrer and allow the batch to settle for about half an hour at 95°C.

Siphon off the clear upper layer of Naphtha extract of Hydrazo compound into the Conversion Kettle as above. Similarly, again extract the Hydrazo compound still remaining in the zinc sludge with a second lot of

250 L of fresh Solvent Naphtha in exactly the same manner as above.

Separate this final naphtha extract of Hydrazo compound and retain it for using in the extraction of the second batch.

3.3 Conversion to Tolidine and Sulfation

Tool A 2500 L capacity M. S. Lead-bonded or fibre-glass lined Kettle with an M. S. jacket and equipped with an Anchor-type lead-lined M. S. Agitator driven by a 3 H. P., 3-Phase, closed Flame-proof Electric motor connected with a Reduction-gear to give 40/30 r. p. m. speeds to the agitator.

The Kettle is further provided with two diametrically opposite M. S. lead-bonded baffles, a manhole, a temperature-recorder, a pressure-gauge, a charging line, two Observation glasses, a horizontal Shell and tube type Distillation Condensor and a bottom-outlet line.

In the clean Kettle, charge

50 L of water.

Start agitator and add

6 Kgs of as is 90% Sodium Bisulfite (M.Wt. 104.0).

Stir to solution.

Heat to 45°C with jacket steam.

Then receive into this the naphtha solution and extract of Hydrazo-Toluene from the above step. This is about 1000 L.

Then charge in it during about 15-20 minutes

440 L of as is 25%w/vol Sulfuric acid solution corresponding to

110 Kgs or real Sulfuric Acid (M. Wt. 98.0) holding the temperature below 50°C.

At the end of addition of the H_2SO_4 solution, run in about

400 L of water of 55-60°C.

Adjust the temperature of the batch in the range of 60-65°C and stir for 2 hours at this temperature.

White coloured o-Tolidine Sulfate falls out of solution.

The re-arrangement and sulfation is complete by now. Stop heating.

Drain off steam from the Kettle jacket.

Circulate cold water through the jacket and cool the batch to 20°C during about 2 hours.

Measure volume of the batch; it will be about

1900 L at 20 C.

Draw a sample of the batch and confirm that precipitation of o-Tolidine sulfate, both from the solvent and aqueous medium, is complete with the following test.

Test Filter a 500 ml batch sample through a Buchner funnel. Separate the upper Solvent layer and the lower aqueous layer from the filtrate. Cool the separated solvent layer as well as the aqueous layer containing Sodium Sulfate, further to 10°C externally. Observe if there is any precipitate of Tolidine sulfate coming out on cooling. It is likely that while cooling the aqueous layer to 10°C, crystals of Sodium sulfate may fall out which should be checked not to confound with Tolidine sulfate.

If both the above layers show no further precipitation of Tolidine sulfate, the precipitation of Tolidine sulfate in the batch is considered to be complete.

When the precipitation of o-Tolidine sulfate is confirmed to be complete, filter the batch through Centrifuge at 20°C.

Transfer the total filtrates to a Storage Vessel for separation and recovery of Solvent Naphtha.

Wash the centrifuge cakes with

200 L of cold water 5-8°C.

Discard the washings to drains.

Suck the cakes well.

Dump the wet cakes of Tolidine sulfate from the Centrifuge, weigh and record.

Weight of wet cakes of Tolidine sulfate,

= Approx. 180-185 Kgs.

3.4. Basification and Filtration

Tool : A 2000 L capacity Wooden vat equipped with a Paddle-type Wooden Agitator driven by a 3 H. P., 3-Phase, closed Electric motor connected with a Reduction-gear to give 40-30 r. p. m. speed to the agitator.

The vat is further equipped with a temperature-recorder, a top Wooden cover with a charging-door on it and a bottom-outlet line leading to the Centrifuge.

		In the clean Vat, charge
1000	L	of water.
		Start agitator.
		Then charge in it the wet centrifuged cakes of Tolidine sulfate.
		Stir for 1 hour without temperature control. Add
0.125	Kg	of Phenyl-Betanaphthylamine dissolved in
1	L	of Acetone.
		Then gradually add over about 1 hour, controlling the foaming, as required about
250	L	of as is 20%w/v Soda Ash solution corresponding to 50 Kgs of real Soda Ash until no test on MRY papers and just no test on B. Y. test papers is obtained. Measure pH of the batch; it should be in the range of 6.7-7.0. If lower, adjust the pH in this range by adding more Soda Ash solution as required. Stir the precipitated Tolidine Base for half an hour without temperature control.
		Measure volume of the batch; it will be about
1350	L	at Room temperature.
		Then add
0.5	Kg	of Sodium Hydrosulfite.
		Stir for 10 minutes.
		Filter the batch through Centrifuge.
		Wash the centrifuge cakes with
300	L	of cold water of 5-8°C.
		Suck the cakes well.
		Dump the base wet cakes from the Centrifuge, weigh and record.
		Analyse a representative sample of the Base wet cake for real o-Tolidine content with a Nitrite titre. Calculate yield.
		Weight of wet cakes of o-Tolidine Base,
		= Approx. 125-140 Kgs.
		= 100 Kgs of real o-Tolidine Base (M. Wt. 212.0)
		= 80% of Theory on o-Nitro-Toluene charge.

4. Raw Materials Usage Data:

	Names of Raw Materials	*M. Wt.*	*Quantity Real (Kgs)*	*Moles*	*One Mole Ratio*
i)	o-Nitro-Toluene	137.0	160.0	1.1678	1.0000
ii)	Zinc dust, as is 90%	–	345.0	–	–
		65.3	310.5	4.7549	14.0717
iii)	Caustic Soda, as is 50%	–	19.0	–	–
		40.0	9.5	0.2375	0.2034
iv)	Sulfuric Acid, 98%	–	112.2	–	–
		98.0	110.0	1.1224	0.9611
v)	Soda Ash, real	106.0	50.0	0.4717	0.4039
vi)	Sodium Bisulfite, as is 90%	104.0	6.0	–	–
vii)	Phenyl-Betanaphthylamine	–	0.125	–	–
viii)	Acetone	–	1.0	–	–
ix)	Solvent Naphtha	Employed	: 1100 L		
		Lost	: 200 L		
		Recovered	: 900 L		
x)	Denatured Alcohol	Employed	: 120 L		
		Lost	: 20 L		
		Recovered	: 100 L		

Purification of o-Tolidine Base:

1. Dissolution and Filtration

In a suitable Wooden Vat,

Charge

1500 L of treated water.

Start agitator and inject a passage of direct steam. Heat to 85°C.

Then charge

200 Kgs of real o-Tolidine Base (M. Wt. 212.0) as a crude product.

Stir for 15-20 minutes at 82°C.

Test for an orange test on B. Y. papers; if not, add a small quantity of 40/50% Caustic Soda lye to obtain a distinct orange test on B. Y. papers. Then add about

1.5-2.0 Kgs of Sodium Hydrosulfite solid and stir at 85°C for 15-20 minutes.

Then charge

250 Kgs of as is 30%w/w HCl corresponding to

75 Kgs of real HCl (Includes a 10% excess over the theoret requirement).

Confirm for a positive blue test on Congo Red papers; if not, add a correct quantity of HCl to get the test. Stir at 80-85°C for about half an hour and confirm that the base has completely gone into solution.

In case if the hydrochloride tends to fall out, increase the dilution with water. Usually, this is not necessary.

When the solution is confirmed to be complete, add to the charge,

10 Kgs of Activated Carbon.

Stir the charge for 45 minutes at 80-85°C.

Filter the batch hot at 80°C through an M. S. R. L. Nutsche Filter dressed with a Terylene filter bag.

Wash the carbon cake in the Nutsche with about

500 L of hot water of 70-75°C.

Receive he washings along with the batch liquor.

Transfer the M. L. and washings to a suitable Wooden Precipitation Vat.

2. Precipitation of Pure Base

Measure volume and temperature of the clarified batch liquor.

Adjust volume to

2500 L at 40°C.

Then add gradually as required under continuous testing, 50%w/w Caustic Soda lye to obtain a just faint-blue test on Congo Red papers.

Note : Do not add excessive lye to effect an increase degree of precipitation during the early stage of addition.

Now add

0.25 Kg of SAD-1 powder dissolved in about 5 L of water (250 gms)

(250 gms) in a bucket.

Then follow an addition of

0.5 Kg of SCA-1 powder dissolved in about 8 L of water

(500 gms) in a bucket.

Stir the batch for 15-20 minutes and draw a sample to observe the colour-change.

Then continue further the addition of 50%w/w Caustic soda lye, as required, to obtain a just faint-orange test on B. Y. papers.

Stir for about 10 minutes and test again on B. Y. papers. The orange test is likely to disappear. Now add to the batch, as required, 20% Soda Ash solution to obtain a permanent faint-orange test on B. Y. papers.

Stir the batch for 1 hour without temperature control. Check for a positive faint-orange test on B. Y. papers.

If this test is negative, further add a corrective quantity of 20% Soda Ash solution to get the test. The precipitation of the base is now complete.

For completion of precipitation, about

152 Kgs of as is 50%w/w Caustic Soda lye corresponding to

76 Kgs of real Caustic Soda may be required.

Filter the precipitated base through a Hydroextractor clothes with a cotton bag.

Wash the product cake in the Hydroextractor with enough volume of ordinary water to no orange test on B. Y. papers.

Suck the product well.

Weigh the wet cakes and record.

Determine the purity of the wet product with a Nitrite titre.

Calculate the nett recover.

Use the evaluated Wet product as such.

14

Manufacture of 1-Phenyl-3-Methyl-5-Pyrazolone

1. Principle:

Aniline oil is diazotized with Sodium Nitrite and Hydochloric Acid. The resulting diazo is reduced to Phenyl Hydrazine Disulfonate with a neutral Sulfite solution and subsequently decomposed to Phenyl Hydrazine Hydrochloride with conc. Hydrochloric Acid. The Phenyl Hydrazine Hydrochloride is condensed with Ethyl-Aceto Acetate and cyclised to 1-Phenyl-3-Methyl-5-Pyrazolone in alkaline medium. Finally, the product is precipitated by controlled neutralization with Hydrochloric acid and isolated by filtration.

2. Reactions:

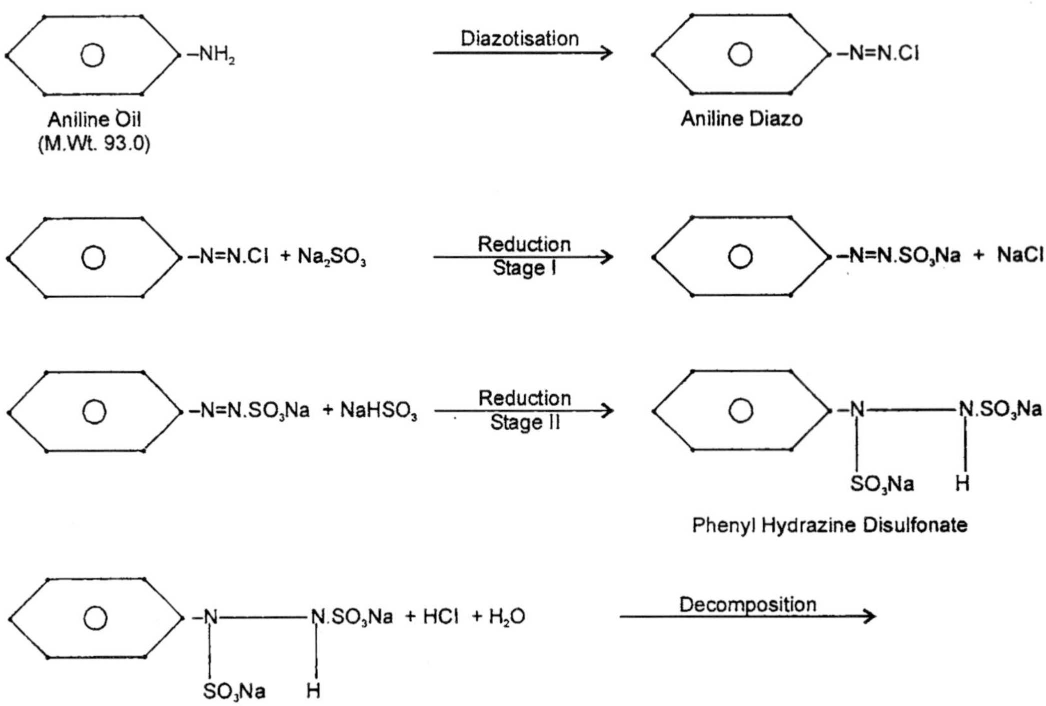

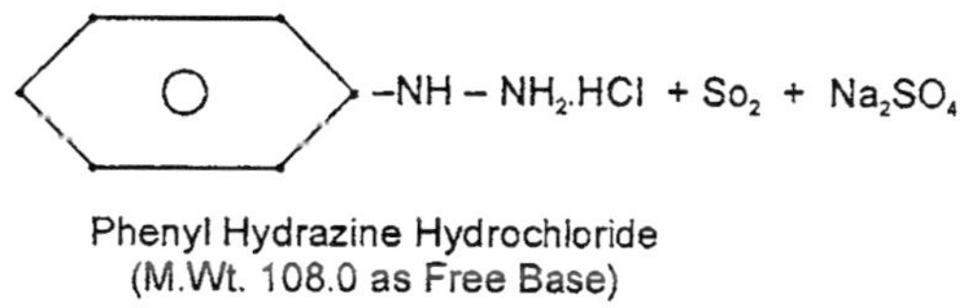

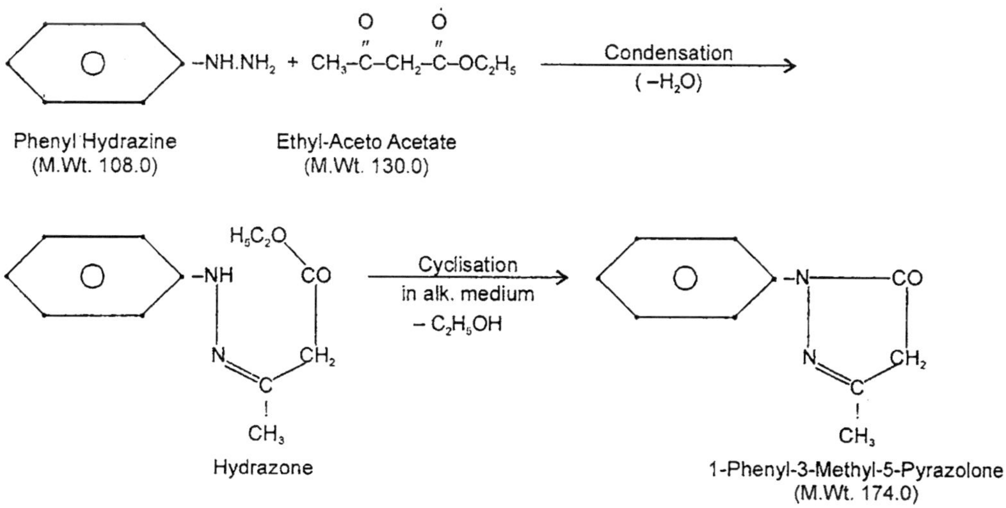

3. Process-Procedure:

3.1. Diazotization of Aniline

Tool A 300 L capacity Wooden Vat with having studded four wooden baffles on the inside wall and equipped with a Paddle-type wooden Agitator driven by a 3 H. P., 3-Phase closed Electric motor connected with a Reduction-gear to give 60 r. p. m. speed to the agitator. The Vat is further equipped with a Temperature-recorder and a bottom-outlet line leading to the Reduction Vessel.

Check tool.

In the clean Diazotization Vat

Charge

450 Kgs of as is 30%w/w Hydrochloric Acid corresponding to

135 Kgs of real HCl (M. Wt. 36.5).

Start agitator and add about

600 Kgs of crushed ice to cool the acid to 0°C.

Then charge gradually under stirring

154 Kgs of real Aniline Oil (M. Wt. 93.0), controlling the fuming. This about takes about 15-20 minutes.

If necessary, add more ice to keep the temperature at 0°C while charging Aniline. At the end of charging, stir the hydrochloride for about

5-10 minutes at 0 to mins 3°C. Add extra ice which keeps floating on the surface of Aniline Hydrochloride charge.

Then add as fast as absorbed, as indicated by tests on KI-Starch papers.

232 L of as is 50%w/vol Sodium Nitrite solution corresponding to

116 Kgs of real Sodium Nitrite (M. Wt. 69.0).

maintaining the diazotization temperature strictly below 3°C by adding requisite quantities of extra ice as required.

During the nitrite addition, the diazotization charge must show a distinct blue test for sufficient acidity on Congo Red test papers. If the test is faint-blue, corrective quantity of 30% HCl must be added at once. The diazotization is treated as complete when a distinct blue test on KI-Starch papers, capable of lasting for 15 minutes, is obtained.

Measure volume of the Diazo, it will be about

2000 L at 0°C.

3.2. Reduction

Tool A 6000 L capacity Wooden Vat having four wooden baffles studded on the inside wall and equipped with a paddle-type wooden Agitator driven by a 5 H. P., 3-Phase, closed Electric motor connected with a Reduction-gear to give 60 r. p. m. speed to the agitator. The vat is further provided with a Temperature-recorder, a wooden top cover bearing an exhaust fan under the hood of the exhaust line leading to the SO_2 scrubbing system and a bottom-outlet valve with the line leading to the Clarifier Nutsche.

(i) Neutral Sulfite Solution

Check tool.

In the clean Reduction Vat,

Charge

600 L of treated water.

Start agitator and charge about

440 Kgs of as is 90-92% Sodium Bisulfite corresponding

397 Kgs of real Sodium Bisulfite (M. Wt. 104.0).

Then add, as required, about

330 Kgs of as is 40%w/w Caustic Soda lye corresponding to 132 Kgs of real Caustic Soda, so as to obtain no test on MRY test papers but just a faint-orange test on B. Y. papers.

Note the temperature of the solution and adjust to 35°C by careful steaming if necessary. Measure volume of the solution and adjust it to

1200 L at 35°C.

Confirm faint-orange test on B. Y. papers.

The pH of the solution will be in the range of 6.9-7.2.

(ii) Reduction

To the neutral Sulfite solution maintained at 35-40°C, add the Aniline diazo solution from Step 3.1, during about 1½ hours maintaining a positive orange test on B. Y. papers but no test on Phenolphthalein papers by making a simultaneous addition of 30%w/vol Sodium Bisulfite solution.

During the Diazo addition, the temperature tends to rise upto 40°C and the pH of the batch is not allowed to fall below 7.0, Since this brings about a brisk decomposition of the diazo and the bisulfite compound. This requires about

100 L of 30%w/vol Sodium Bisulfite solution corresponding to

30 Kgs of real Sodium Bisulfite (M. Wt. 104.0).

The rising alkalinity during the Diazo addition is checked as above by the addition of Sodium Bisulfite.

At the end of Diazo addition, the pH of the reduction mass will be about 7.1-7.3 which tends to rise further. At the end of Diazo addition, stir the batch for half an hour at 35-40°C, frequently checking the pH for the increasing alkalinity.

During this stirring (at the end of Diazo addition), control the rising alkalinity by adding 1: 1 Solution of Hydrochoric Acid, as required, to maintain a positive orange test on B. Y. papers but no test on Phenolphthalein papers.

At the end of 1/2 hour stirring at 35-38°C slowly steam the batch to 80-85°C over a period of about 2 hours. During this period also, control the fast rising alkalinity by adding to the batch, as required, 1: 1 Hydrochloric Acid solution to conform to the above tests.

For this pH adjustment, about

110 Kgs of as is 30%w/w HCl corresponding to

33 Kgs of real HCl will be required.

With the completion of this addition, the rising pH nearly stabilizes and becomes steady in the range of 7.2-7.5.

The pH range for ideal reduction is 7.2-7.5.

Stir the reduction at 80-85°C and in the range of pH 7.2-7.5 until the reduction is complete.

Disappearance of yellow colour of the reduction mass denotes the completion of reduction.

This takes about 4-1/2 to 5 hours.

At the end of completion of reduction, a little coloured material or particles are seen in suspension. Decolourise this coloured material by adding to the batch about.

1-2 Kgs of Sodium Hydrosulfite.

The resulting batch solution is almost colourless.

Measure volume of the batch: it will be about

4000 L at 80-85°C.

3.3 Hydrochloridation

Close the top wooden cover of the Vat holding the reduced batch liquor.

Open the exhaust line leading to the SO_2 Gas Scrubbing System.

Start the exhaust fan under the hood of the exhaust line.

To the reduced batch solution stirring at 80-85°C. Add gradually and carefully over about 1-1/2 hours

410 Kgs of as is 30%w/w Hydrochloric Acid equivalent to

123 Kgs of real HCl, controlling the brisk evolution of SO_2 gas.

Confirm a strong blue test on the Congo Red test papers at the end of addition of HCl.

This acid addition decomposes the Phenyl Hydrazine Sulfonate to Phenyl Hydrazine which combines with excess of Hydrochloride.

Stir the batch at 90-92°C until SO_2 gas is quantitatively driven out from the system. This takes about 1-1/2 to 2 hours after the end of HCl addition.

No decolourisation of moistened blue KI-Starch indicator papers denotes the absence of SO_2 gas. The Phenyl Hydrazine Hydrochloride tends to crystallize out at or below 70°C and therefore, the temperature of the batch should not be allowed to fall below 75°C.

The hydrochloridation and removal of SO_2 gas is complete by now.

The Phenyl Hydrazine Hydrochloridg is in complete solution.

To the clear Hydrochloride solution add

5 Kgs of Active Carbon and

2.5 Kgs of Hyflosupercel.

Stir for ½ hour at 80-85°C.

Measure volume of the batch: it will be about

4700 L at 85-90°C.

Filter the batch hot at 80-85°C

through a Clarifier Pres or Wooden Nutsche. Transfer the clear colourless batch filtrate to the Condensation Vessel described below.

Wash the Reduction Vat with about

100 L of hot water of 80°C and pass this washing over the cake in the Clarifier.

Collect this filtered washing in the Condensation Vessel holding the main batch filtrate.

Discard the filter cake to drains.

3.5. Condensation and Cyclisation

Tool : A 7500 L capacity Wooden Vat with having studded four wooden baffles on the inside wall and equipped with a Paddle-type wooden agitator driven by a 5 H. P., 3-Phase, closed Electric motor connected with a Reduction-gear to give 40 r. p. m. speed to the agitator. The Vat is further equipped with a Temperature-recorder and a bottom-outlet line leading to the Clarifier.

3.5. (i) Condensation:

Check tool.

Receive the batch filtrate along with washing from the above step into the clean Condensation Vat.

Measure volume; it will be about

500 L at 68-70°C.

Start agitator and steam the batch to 75-80°C.

Add to this as required, about

440 Kgs of as is 40%w/w Caustic Soda lye corresponding to

176 Kgs of real Caustic Soda to remove the blue test on Congo Red papers.

Then add as fast as possible

215.0 Kgs of real Ethyl-Aceto Acetate (M. Wt. 130.0).

A white turbulent solution with a greenish tings is obtained.

Check for a distinct blue test on Congo Red papers which is developed as soon as the ester is added up. Then immediately follow the addition of 40%/w/w Caustic Soda lye to obtain a just faint-red test on B. Y. papers. For this, about

175 Kgs of as is 40%w/w Caustic Soda lye corresponding to

70 Kgs of real Caustic Soda will be required.

Stir for one hour at 75-80°C maintaining a faint but distinct orange test on B. Y. papers.

The pH of the batch will be in the range of 7.2-7.5. Normally, the faint B. Y. test once obtained remains steady during the one hour agitation period.

Test for the completion Condensation as follows:

Test : Spot a drop of the batch sample on a piece of filter paper. Cross-spot it with a drop of 0.5% ABG Indicator solution. If any unconverted Phenyl Hydrazine is present, the indicator dye will produce a violet colour in the spot. If no violet colour is produced, the condensation is treated as complete.

ABG Indicator is a dye obtained by coupling diazotized *p*-Amino-Benzaldehyde with G-Salt.

The formation of Phenyl Hydrazon is complete by now.

3.5. (ii) Cyclisation:

Now add to the batch as is 40%w/w Caustic Soda lye to obtain a distinct red test on Mimosa

Indicator papers.

Keep the batch stirring at 80-85°C and go on adding 40%w/w Caustic Soda lye as fast as consumed, as indicated by the fading test on Mimosa papers. Quickly obtain a distinct red test on Mimosa papers as soon as the test disappears or fades down, with the addition of 40%w/w Caustic Soda lye.

Finally, on continued addition of 40%w/w Caustic Soda lye the distinct red test on Mimosa stabilises at one stage.

This indicates the sufficiency of Caustic Soda addition. Stir the batch at 80-85°C for 2 hours after the distinct red test on Mimosa is stablised.

During this agitation period, frequently check for a strong red test on Mimosa.

If the test still fades down, further add a corrective quantity of 40%w/w Caustic Soda to get the distinct test. A permanent and distinct red test lasting for full two hours on Mimosa papers indicates the completion of cyclisation.

The addition of Caustic soda takes about 2 hours and the total requirement corresponds to about

235 Kgs of as is 40%w/w Caustic Soda lye equivalent to

94 Kgs of real Caustic Soda.

At the end of cyclisation, the colour of the batch solution is pale-yellowish-green.

The solution is clear and shows no material out of solution.

Then slowly add to the batch in a thin stream as required about

80 Kgs of as is 30%w/w HCl equivalent to

24 Kgs of real HCl to remove and test on Mimosa papers and to reduce to a faint-pink but positive test on Phenolphthalein papers.

Note pH of the batch liquor; it will be in the range of 8.3-8.6.

Then add

25 Kgs of Active Carbon and

5 Kgs of Hyflosupercel.

Stir the batch for 1 hour at 75-80°C.

Measure volume of the batch; it will be about

5800 L at 75-80°C.

Filter the batch hot at 75-80°C through a Clarifier Press or Nutsche Filter.

Collect the clear batch filtrate in the Precipitation Vessel described below.

Rinse the inside wall of the Condensation Vat with about

100 L of hot water of 70-75°C, stir for a few minutes and pass on the cake in the Clarifier.

Collect the clarified washing along with the main batch filtrate in the Precipitation Vat.

Discard the Clarifier cakes to drains.

The clarified batch liquor is almost colourless or at the most a pale-reddish-amber.

The pH of the batch liquor is about 8.2-8.5.

3.6 Precipitation

Tool A 10,000 L capacity Wooden Vat with having studded four wooden baffles on the inside wall and equipped with a paddle-type wooden Agitator driven by a 5 H. P., 3-Phase, closed Electric motor connected with a Reduction-gear to give 40 r. p. m. speed to the agitator. The Vat is further equipped with a portable Cooling coil, a Temperature-recorder and a bottom-outlet line leading to the Nutsche Filter.

Check tool.

Receive the clarified batch liquor along with the washing in the clean precipitation Vat.

Measure volume; it will b about

6000 L at 65-70°C.

Start agitator

Cool the batch liquor to 40°C by circulating chilled water through the Cooling coil and adjust the volume to

7000 L at 40°C.

Then add while stirring as required, about

240 Kgs of as is 30%w/w Hydrochloric Acid corresponding to

72 Kgs of real HCl to obtain a just faint-violet test on Congo Red papers.

The addition of HCl takes about 1-1-1/2 hours.

Stir for 10-15 minutes and check the faint-violet Congo test.

Then remove the Congo Test by adding to it, as required, about

10 Kgs of Sodium Acetate Crystals.

Note pH of the batch.

Adjust the pH in the range of 5.5-6.0 by adding necessary quantity of Sodium Acetate crystals or Sodium Bicarbonate.

The product precipitates out in the form of fine buff to pale-grey coloured coarse particles. Finally, cool the batch at 28-30°C and stir for ½ hour at this temp.

Filter the batch at 28-30°C through a Nutsche Filter.

Rinse the Precipitation Vat with about

100 L of cold water of 25°C, stir for few minutes and pump it over the product cake in the Nutsche. Suck the product well.

Discard the M. L. and washing to drains.

Dump the Nutsche.

Weigh the wet product and record.

Dry the wet product in air-dryers at 80-85°C until dry.

Weigh the dry product and record weight.

Grind the dry product.

Analyse a sample of the ground product with a coupling titre for real content and calculate yield.

4. Yield:

Weight of wet product,

= Approx. 325-350 Kgs.

= Approx. 250 Kgs of dry, 1-Phenyl-3-Methyl-5-Pyrazolom containing 98% real material on M. Wt. 174.0.

= 245 Kgs of real o-Phenyl-3-Methyl-5-Phrazolone containing 98% real material on M. Wt. 174.0.

= 245 Kgs of real o-Phenyl-3-Methyl-5-Pyrazolone

(M. Wt. 174.0)

= 85% of Theory on Aniline charge.

5. Raw Materials Usage Data

	Name of Raw Material	*M. Wt.*	*Quantity Real (Kgs)*	*Moles*	*On Mole Ratio*
1.	Aniline Oil	93.0	154.0	1.6559	1.0000
2.	HCl, real	36.5	135.0	3.6986	2.2336
3.	Sodium Nitrite	69.0	116.0	1.6811	1.0152
4.	Sodium Bisulfite, real	104.0	397.0	3.8173	2.3052
5.	Caustic Soda, real	40.0	132.0	3.3000	1.9928
6.	Sodium Bisulfite, real	104.0	30.0	0.2884	0.1742
7.	HCl, real	36.5	33.0	0.9041	0.5460
8.	Sodium Hydrosulfite	–	2.0	–	–
9.	HCl, real	36.5	123.0	3.3698	2.0350
10.	Active Carbon	–	5.0	–	–
11.	Hyflosupercel	–	2.5	–	–
12.	Caustic Soda, real	40.0	176.0	4.4000	2.6571
13.	Ethyl-Aceto Acetate	130.0	215.0	1.6559	1.0000
14.	Caustic Soda, real	40.0	70.0	1.7500	1.0568
15.	Caustic Soda, real	40.0	94.0	2.3500	1.4191
16.	HCl, real	36.5	24.0	0.6575	0.3970
17.	Active Carbon	–	25.0	–	–
18.	Hyflosupercel	–	5.0	–	–
19.	HCl, real	36.5	72.0	1.9726	1.1912
20.	Sodium Acetate, Cryst.	136.0	10.0	0.075	0.0444

Process-Procedure

1. Dissolution of Crude PMP:

		Charge
4000	L	of water in a 7000 L capacity Wooden Vat.
		Start agitator and charge
500	Kgs	of ordinary Common Salt.
		Pass live steam in the charge and heat to 60°C.
		Then charge
300	Kgs	of as is 97-98% Crude Pyrazolone.
		Then charge, as required, about
152	Kgs	of as is 50%w/w Caustic soda lye equivalent to

76	Kgs	of real Caustic soda, so as to obtain a distinct red test on Phenolphthalein papers.
		Stir the PMP to solution.
		Draw a sample in a 2 L beaker after 15-20 minutes and confirm that the PMP is in complete solution.
		The solution of PMP is dark-amber in colour.
		Heat the solution further to 80°C with live steam.
		Then charge quickly
12	Kgs	of Sodium Hydrosulfite powder.
		Stir for 10 minutes at 75-80°C.
		Draw about 2 litres of sample in a 3 L beaker and observe the colour of the solution. The solution shows a pale-yellow colour.
		Then charge
20	Kgs	of Activated Carbon.
		Stir the batch for one hour without temperature control.
		Filter the batch through the Nutsche Filter and wash the Carbon cakes with about 200 L of hot water.
		Transfer the clear filterate and washings in the clean Precipitation Vat.

2. Precipitation of pure PMP

		Measure volume of the filtered batch solution and adjust it to
5400	L	at 45°C by adding ice and water as required. Start agitator and add
3	Kgs	of Sodium Hydrosulfite powder.
		Stir for 15 minutes.
		Then add to it a solution of
3	Kgs	UDA, dissolved in
25	L	of ordinary water.
		Then start the gradual addition of 30% HCl through the Syphon tube.
		During the HCl addition, frequently test on B. Y. papers and serve the HCL quantity enough to just remove orange test on B. Y. papers.
		At once stop the HCl addition when the test on B.Y. papers just becomes negative.
		For completing the precipitation, the quantity of HCl required will be about
180	Kgs	of as is 30%w/w HCl, equivalent to
54	Kgs	of real HCl.

Stir for 15-20 minutes and filter the precipitated product through the Nutsche Filter.

Wash the product with about

400 L of ice-cold water in the Nutsche.

Suck well and dump the we product.

Dry the wet product in air-dryers at 80-85°C.

Weigh the pure, dry product.

3. Yield:

Weight of the pure, dry PMP,

= About 245-250 Kgs, having a purity of about 98.5-99.%

= 83% on the charge of Crude product.

15

Manufacture of p-Nitro-Aniline-o-Sulfonic Acid, NH_4 Salt

1. Principle:

4-Chloro-Nitrobenzene-3-Sulfonic Acid is aminated with Ammonia solution under pressure, the excess ammonia distilled off and the resulting product is finally cooled and isolated by filtration. The nutsche cakes are centrifuged and the wet centrifuged product is evaluated for real content.

2. Reactions:

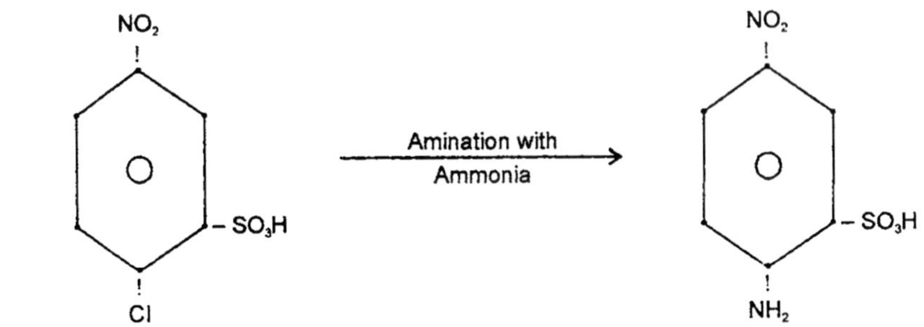

4-Chloro-Nitrobenzene-3-Sulfonic Acid (M.Wt. 237.5)

p-Nitro-Aniline-o-Sulfonic Acid (M.Wt. 218.0)

3. Process-Procedure:

3.1. Amination

Tool A 2500 L capacity Special M S. Autoclave with an M. S. Limpet Coil and Jacket and tested for 250 p. s. i. working Pressure. It is equipped with a Gate-type special M. S. agitator driven by a 5 H. P., 3-Phase, closed Electric motor connected with a Reduction-gear to give 40 r. p. m. speed to the agitator. The Autoclave is further provided with a manhole, a Pressure-gauge, a Temperature-recorder, an Ammonia distillation and absorption Equipment and a blow-out line leading to the Isolation Vessel.

Check tool.

In the clean autoclave, charge

1020 L of as is 26%w/vol Liquor Ammonia corresponding to

265 Kgs of real Ammonia (M. Wt. 17.0).

Star agitator.

Charge into it the centrifuged wet cakes of 4-Chloro-Nitrobenzene-3-sulfonic Acid,

Corresponding to a nett charge of

585 Kgs of real 4-Chloro-Nitrobenzene-3-sulfoni Acid (M. Wt. 237.5). This charging takes about 1-1/2 hours.

Stir the charge about 1 hour to obtain a smooth slurry. Check volume; it will be about

1500 L.

Then finally adjust the volume of the charge to

1600 L using 26%w/vol Liquor Ammonia.

This volume adjustment requires about

100-150 L of as is 26%w/vol Liquor Ammonia.

Close the Autoclave.

With a passage of steam in the limpet coil, heat the batch to 120°C during 4 hours; then shut off steam. Allow the temperature to rise of its own to 145 degC. This takes about 1-1/2 to 2 hours.

Note : While the temperature is rising of its own, take care when it reaches to 140°C. To prevent overshooting the temperature limit of 145°C, control the rising temperature by circulating cold water through the jacket for a while until the rise comes under the check.

Adjust the batch temperature at 144-146°C.

The pressure developed at this is expected to be in the range of 144-146°C for 5-1/2 hours.

While nearing the end of the reaction period, the pressure tends to fall by a value of about 10-15 p.s.i. (But in actual practice, the pressure is about 120-110 p.s.i.)

Note : The temperature limit is very important here for a good reaction.

When the reaction period is over, shut off steam and cool the batch to 100°C by circulating cold water through the jacket.

Then stop circulation of cold water and gradually open the valve leading to the Ammonia distillation and absorption system. The residual pressure in the Autoclave at 100°C is about 60 p.s.i. which is gradually released with the evolution and absorption of ammonia.

When the pressure in the Autoclave falls to the level of atmospheric pressure, discontinue ammonia distillation.

The amidation is complete and the batch is now ready for isolation.

3.2. Isolation

Tool A 2500 L capacity M. S. Kettle with an M. S. jacket and equipped with an Anchor-type agitator driven by a 5 H. P., 3-Phase, closed Electric motor connected with a Reduction-gear to give 40 r. p. m. speed to the agitator.

The Kettle is further equipped with a charging-line, a vent-line opening to the atmosphere, a temperature-recorder and a bottom-outlet line leading to the pressure-nutsche.

Check tool.

Receive the batch from the above step into the clean Isolation Kettle and start agitator.

Circulate ordinary water through the jacket to cool the batch to 35°C. Then circulate chilled water to further cool the batch to 25°C. Stir for half an hour at 25°C.

The batch is now ready for filtration.

Blow the batch to Pressure-Nutsche.

Suck the M. L.

Rinse the Isolation Kettle with about

200 L of the M. L., stir and transfer to the nutsche. Suck the M. L. and discard it to drains. Dump the wet product from the nutsche. Centrifuge the nutsche cakes. Dump the centrifuged product, weigh and record the weight.

Draw a representative sample of the wet centrifuged product and determine its real content as

p-Nitro-Aniline-o-Sulfonic Acid (M. Wt. 218.0) with a Nitrite titre.

Calculate yield.

The final wet product is as Ammonium Salt of *p*-Nitro-Aniline-o-Sulfonic Acid.

4. Yield:

Weight of centrifuged Wet Cakes,

= Approx. 575-600 Kgs.

= 500 Kgs of Real *p*-Nitro-Aniline-*o*-Sulfonic Acid (M. Wt. 218.0)

= 93% Theory yield on 4-Chloro-Nitrobenzene-3-Sulfonic Acid charge.

Recover:

150 Kgs of real Ammonia.

5. Raw Materials Usage Data:

	Names of Raw Materials	*M. Wt.*	*Quantity Real(Kgs)*	*Moles*	*One Mole Ratio*
i)	4-Chloro-Nitrobenzene-3-Sulfonic Acid	237.5	585.0	2.4631	1.0000
ii)	Liquor Ammonia, 26%w/v	–	1150 L		
	containing Real NH_3	17.0	300.0	17.6470	7.1645

16

Manufacture of 2-Naphthylamine-3:6:8-Trisulfonic Acid, Disodium Potassium Salt Paste

1. Principle:

Dry Amido–G-Acid is sulfonated with 65% Oleum under controlled conditions. The sulfonated batch is drowned into an ice-water mixture and finally isolated as Potassium Salt. The product is filtered through Nutsche Filters. The wet product is evaluated for real content and is used or sold as such in wet form.

2. Reactions:

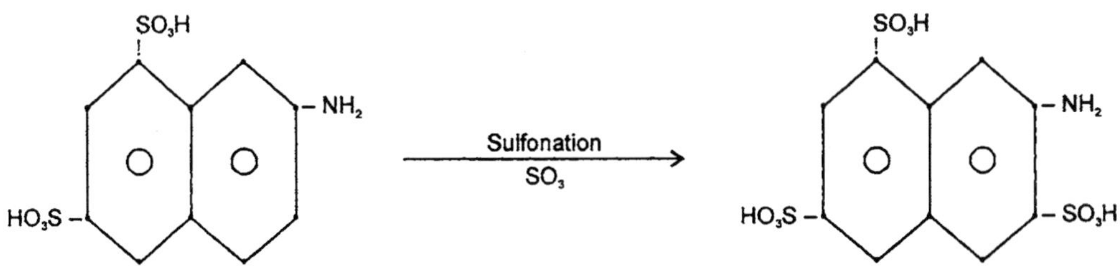

Amido-G-Acid
(M.Wt. 303)

2-Naphthylamine-3:6:8-
Trisulfonic Acid
(M.Wt. 383)

3. Process-Procedure:

3.1. Sulfonation

Tool : A 2,000 L capacity, C. I., jacketed Sulfonation Kettle equipped with a C. I. Anchor-type Agitator driven by a 5 H. P., 3-Phase, closed Electric motor connected with a Reduction-gear to give 40 r. p. m. speed to the agitator. The Kettle is further equipped with a Temperature-Recorder-cum-Control, a

Pressure-gauge, a Vent-line leading to the Scrubber via a Fume-duct, a manhole, and a blow-out line leading to the Drowning Vessel.

Check tool.

In the clean, dry Sulfonation Kettle,

Charge

425.0 Kgs of Monohydrae, i.e. 100% Sulfuric Acid.

Start agitator.

Then Charge

305.0 Kgs of 65% Oleum, together with a corrective additional quantity of Oleum, enough to bind up the total moisture content in the charge of Amido-G-Acid, to follow thereafter.

N. B. : For this, the moisture content of an Amido-G-Acid sample should be predetermined.

Every 18 Kg weight of water (as moisture) requires 80 Kgs of real Sulfur Trioxide to bind it to Monohydrate. From this, calculate the exact quantity of 65% Oleum to be added. Record this quantity of Oleum.

This operation takes about 45 minutes.

Then charge while stirring, allowing the temp to rise up to 85-90°C during charging,

465.0 Kgs of Real Amido-G-Acid (M. Wt. 303), taking about 1-1/2 hours. The Amido-G-Acid used is a powder of about 10-15 Mesh. At the end of charging, the mass becomes thick. Add some 65% Oleum to thin the batch and record the quantity of Oleum added.

Stir the batch for 20 minutes.

Then charge in it at 92-95°C,

825.0% Kgs of 65% Oleum, deducting from this the quantity of Oleum already used for thinning the batch, slowly during ablout 3 hours.

Heat the batch to 130°C slowly over a period of 1½ hours with jacket steam.

Stir the batch for 15 hours at 129-131°C.

N. B. : Take care to avoid excessive loss of SO_3 during the heating period at 130 degC, since this prolongs the sulfonation time. Keep the SO_3 loss to a minimum by adjusting the outlet cock to the vent so that the Sulfonator may be under small positive pressure of not more than 5 p.s.i.

At the end of 15 hours of sulfonation, carefully draw a sample of the sulfonated mass and analyse it for checking the completion of sulfonation. After drawing the sample, allow the sulfonation to proceed at 130 ± 1°C until the analysis is carried out.

Test : **Completion of Sulfonation:**

Measure 50 ml of the sample in a hot, dry 50 ml measuring cylinder. Pour it slowly into a 1 litre beaker containing about 400 gms of crushed ice. Rinse the cylinder with cold water and transfer it to the beaker. Heat the mixture under stirring to 90°C and add 50 gms of pure Sodium Chloridge.

Cool the mixture externally to 25°C and stir it for half an hour at 25°C.

Filter the slurry at 25°C through a Buchner funnel in a clean flask. Press the filter cake and suck dry, but do not wash with water.

Dilute the filtrate to about 1000 ml volume in a 2-litre beaker with ice-cold water. Add 25 ml of Conc. HCl and titrate under stirring with a Standard N/1 Sodium Nitrite solution to a 3-minute end-point on KI-Starch papers. Note the burette titre. The titre should not exceed 3 ml of N/1 Sodium Nitrite solution. If the titre exceeds 3 ml of N/1 sodium Nitrite solution, the sulfonation is deemed to be partially incomplete.

If the sulfonation appears to be incomplete as justed from the above test, stir the sulfonation further for 2 hours at 130 ± 1°C.

Retest a sample as above.

If the titre still exceeds 3 ml after the additional 2 hours of sulfonation, and to the batch.

40.0 Kgs of 65% Oleum and continue sulfonation at 130 ± 1°C for further 2 hours.

Resample the bath and test as above for checking the completion of sulfonation. Usually, the sulfonation is complete by now. In any case, add no further Oleum and treat the sulfonation as complete at this stage. Stop steam and drain off the jacket. Circulate water in the jacket and cool the batch to 50°C.

Volume of the sulfonated batch will be about

1200 L at 50°C.

Total time of sulfonation operation = About 30 hours.

3.2. Drowning and Isolation

Tool : A 7,500 L capacity M. S. R. L. Tile-lined Kettle equipped with a Paddle-type M. S. R. L. Agitator driven by a 7.5 H. P., 3-Phase, closed Electric motor connected with a Reduction-gear to give 30 r. p. m. speed to the agitator. The Kettle is further equipped with a Thermometer, a Vent-line, an immersion Coil of lead, a manhole and a bottom-discharge line leading to the Nutsche Filters.

Check tool.

		In the clean Drowning Kettle,
		charge
400	L	of Treated water.
		Start agitator.
		Charge
450	Kgs	of rushed ice.
		Then slowly blow into this the sulfonated mass from the above step during about 2 hours, allowing the temperature not to exceed 50°C with the addition of requisite quantity of extra ice.
		For this, about
1600	Kgs	of extra ice may be required.
		Stir the drowned mass for about 15 minutes.
		Then heat the drowned batch to 95°C by passing L. P. Steam through the immersion Coil.
		Measure volume and adjust it to
4600	L	at 95°C by adding water as required. Add
3.5	Kgs	of Lubrol E and Stir the batch for 20 minutes to ensure complete dispersion.
		Then add to the batch, evenly over a period of 1½ hour,
235.0	Kgs	of Potassium Sulfate holding the batch temperature in the range of 95 ± 1°C.
		Stir the batch at 95 ± 1°C for 1 hour.
		Finally, cool the batch uniformly over a period of 7 hours to 25°C by circulating ordinary and chilled water in the Coil, as is necessary. The product falls out in a fine crystalline form. Stir the batch further for 1 hour at 23-25°C.
		The isolation is complete.
		Sample the isolated batch and determine the liquor-loss.
		Total time for Isolation = about 20-22 hours.

Test : **Determination of Liquor-loss:**

Filter a measured volume of the isolation sample through a Buchner funnel in a dry, clan Buchner flask. Suck the cake well but do not wash. Measure volume of the filtrate and record. Measure 500 ml of this clear filtrate in a 1-litre beaker. Add to it 25 ml of gonc. HCl. Cool to 10°C by adding enough ice and titrate, under agitation, with a Standard N/1 Sodium Nitrite solution to a 3-minute end-point on KI-Starch papers. Note the Burette reading and call it "T".

Batch Volume = "V" Litres

Volume of isolation sample to obtain 500 ml filtrate = "v" ml.

Then, percentage Liquor-Loss,

$$\frac{T \times N.\text{ of } NaNO_2 \text{ Soln.} \times 0.383 \times V \times 100}{v \times 588}$$

= % of Theory.

The total Liquor-Loss should not be more than

6% of Theory.

3.3 Filtration and Washing

When the test for Liquor-loss is satisfied, run the batch slurry by gravity to the two tile-line Nutsche Filters and suck the Mother liquor into the Vacuum Receivers. Charge in the empty Drowning Kettle

800 L of Treated water.

Stir and charge

40.0 Kgs of Potassium Sulfate.

Stir to solution.

When mot of the Mother liquor is sucked off from the cakes, break the vacuum.

Then run the above prepared Potash wash liquor from the Drowning Kettle, in two equal portions, over the cakes in the two Nutsche Filters and allow the wash-liquor to soak in the cakes for about 1 hour.

Finally, suck the cakes receiving washings into the Vacuum Peceivers.

Suck the cakes as dry as possible.

Discard the M. L. and washings to acid-drains.

Dump the Nutsche Filters.

Wight the wet cakes and record.

Store the wet cakes in wooden cakes or suitable containers. Draw a representative sample of the wet cakes and determine real content.

4. Yield:

Weight of Wet Cakes.

= Approx. 1250-1300 Kgs.

= 500 Kgs of real 2-Naphthylamine-3:6:8-Trisulfonic acid (M. Wt. 383.0)

= 85.% of Theory on the charge of real Amido-G-Acid.

5. Raw Materials Usage Data:

	Names of Raw Materials	*M. Wt.*	*Quantity Real (Kgs)*	*Moles*	*One mole Ratio*
i)	Amino-G-Acid	303.0	465.0	1.5346	1.0000
ii)	Sulfuric Acid, 100%	98.0	425.0	4.3367	2.8259
iii)	Oleum, 65%	–	1130.0	–	–
	SO_3:	80.0	734.5	9.1812	5.9828
iv)	Potassium Sulfate	174.0	275.0	1.5804	1.0298
v)	Lubrol E (Dispersing Agent)	–	3.5	–	–

17

Manufacture of m-Phenylene-Diamine-4-Sulfonic Acid

1. Principle:

Solid m-Phenylene Diamine (MPD), about 98-99%, is sulfonated with 20% Oleum to m-Phenylene-Diamine-4-Sulfonic Acid and partly to some 4:5 and 4:6-Isomers which are subsequently hydrolysed by boiling with weakened Sulfuric Acid to the Monosulfonic Acid, thus resulting in the formation of the objective compound. The hydrolysed batch is finally drowned into an ice-water mixture and the precipitated product is filtered through Nutsche Filters. The wet product is further sucked in Hydroextractor to raise the solid contents. The centrifuged cakes are finally evaluated for real content and the yield calculated. The product is used or sold in the form of wet cakes.

2. Reactions:

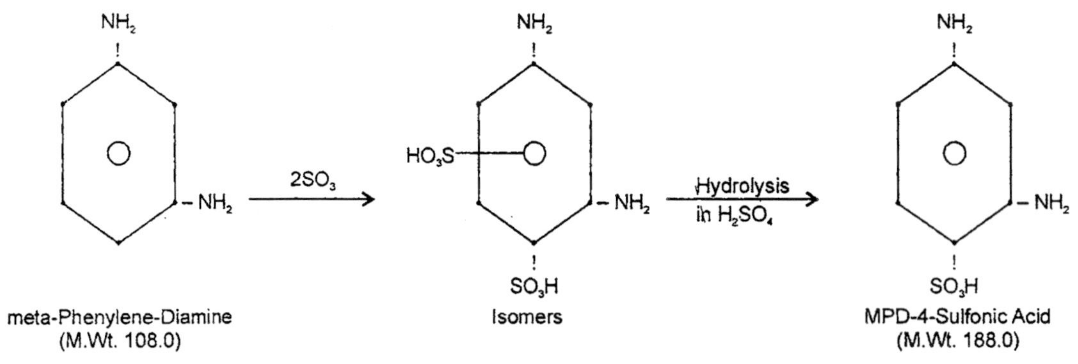

3. Process-Procedure:

3.1. Sulfonation

Tool A 2500 L capacity C. I. jacketed Sulfonation Kettle having an Anchor-type C. I. Agitator driven by a 5 H. P., 3-Phase, closed electric motor connected with a Reduction-gear to give 40 r. p. m. speed to the agitator. The Kettle is further equipped with a Temperature-Recorder-cum-Control, a vertical Vent-line, a manhole, a charge-hole for liquid charges, a Pressure-gauge and a Blow-out line leading to the Drowning Vessel.

Check tool.

In the clan, dry Sulfonator,

Charge

1,070.0	Kgs	of as 98% Sulfuric Aid corresponding to
1,048.6	Kgs	of real Sulfuric Acid.
		Stir and charge in it slowly
380.0	Kgs	of as is 25% Oleum, controlling the fuming.
		The water contained in the 98% Sulfuric Acid is bound up by the Sulfur Trioxide of Oleum resulting in the quantitative formation of Monohydrate. This mixture corresponds to 1,450 Kgs of Monohydrate, i. e. 100% Sulfuric Acid. Note the temperature of Monohydrate and record.
		Start circulation of cooling water in the jacket. Then add to the Monohydrate slowly, under continuous agitation.
336.0	Kgs	of Real dry *m*-Phenylene-Diamine (M. Wt. 108.0) over a period of about 1-1/2 hours, allowing the temp not to exceed 86°C during charging. The MPD employed should be of Fused or Distilled Grade, dry and of good quality. Adjust the temp between 86-88°C by regulating flow of cooling water in the jacket. Then add to the Sulfonation mass
725.0	Kgs	of as is 60% Oleum corresponding to
435.0	Kgs	of Real Sulfur Trioxide (M. Wt. 80) in a thin flow over a period of one hour maintaining the temp of Sulfonation in the range of 86-90°C. Stop circulation of chilled brine.

Drain down brine from the Kettle jacket. Then start a passage of steam in the Kettle jacket and gradually heat the Sulfonation mass to 156°C during about 3½ hours and continue the Sulfonation at 155-157°C for 4½-5 hours.

The sulfonation is complete now positively complete. Stop steam.

Drain down steam from the jacket.

Then cool the batch to 88°C, first by circulating ordinary water and then chilled water, if necessary, in the jacket. This takes about 1/1¼ hours.

3.2. Hydrolysis of the disulfonic Acid Isomers

Now slowly and cautiously, charge in the Sulfonation mass, while stirring,

88.0 L of water in a thin stream during about 1-1/4 hour maintaining the temperature between 88-91°C by jacket-cooling.

Then drain off water from the jacket.

Start a regulated passage of steam in the jacket and heat the batch gradually to 135°C during 2 hours. Maintain the batch at 136-138°C for 2½ hours more.

The hydrolysis is now complete.

Finally, drain off steam from the jacket and cool the batch a 88°C during about 1-1/4 hour by circulating cooling water in the jacket.

3.3 Drowning the Hydrolysed Mass

Tool : A 6,000 L capacity, M. S., Lead-lined jacketed Kettle equipped with a gate-type, M. S., Lead-lined Agitator driven by a 5 H. P., 3-Phase, closed electric motor connected with a Reduction-gear to give 40 r. p. m. speed to the agitator. The Kettle is further equipped with Thermometer, a vertical vent-line, a manhole and a bottom-discharge line leading to the Nutsche Filter.

Check tool.

Charge in the clean Drowning Kettle

600 L of treated water and

2,300 Kgs of crushed Ice.

Start the agitator.

Then blow the Hydrolysed bath from the Sulfonator into the Ice & water mixture at such a rate that the temperature of the final drowning mixture comes between 40-50°C.

Start a flow of ordinary water in the jacket and stir the batch to cool to 30°C during about 2 hours. Then start a regulated flow of chilled water in the jacket and cool the batch to 18°C, gradually over a period of 4 hours.

MPD-4-sulfonic Acid precipitates out in a fine form at the end of this period.

The batch is now ready for filtration.

3.4. Filtration

Filter the Batch cold at 18°C through

Tile-lined Product Nutsches.

Suck the M. L. Well.

Wash the Nutsche cakes with

300 L of as is 25%w/vol Brine.

Finally, suck the cakes well.

Dump the Nutsche.

Discard the M. L. and washings to acid-drains.

Centrifuge the wet cakes to further suck down the liquor. Dump the centrifuge.

Weigh the centrifuged Wet Cakes and record. Draw a representative sample of the centrifuged wet cakes and determine its real content.

Calculate and record yield.

Note : The filtration of the batch may be carried out straight in M. S. R. L. Centrifuge, avoiding the Nutsche filtration. Even brine-washing of the wet Product is convenient in a Centrifuge.

4. Yield:

Weight of Centrifuged Wet Cakes,

= Approx. 100-1100 Kgs containing about 45-50%. Real *m*-Phenylene Diamine-4-Sulfonic Acid (M. Wt. 188.0).

= 500.0 Kgs of real

m-Phenylene-Diamine-4-Sulfonic Acid (M. Wt. 188.0)

= 85.5% of Theory on the charge of real MPD.

5. Raw Materials Usage Data:

	Names of Raw Materials	*M. Wt.*	*Quantity Real (Kgs)*	*Moles*	*One Mole Ratio*
i)	m-Phenylene-Diamine	108.0	336.0	3.1111	1.0000
ii)	Sulfuric Acid	98.0	1048.6	10.7000	3.4393
iii)	Oleum, 25%	-	380.0	–	–
	SO_3 :	80.0	95.0	1.1875	0.3817
iv)	Oleum, 60%	–	725.0	–	–
	SO_3 :	80.0	435.0	5.4375	1.7478
v)	Common Salt in Brine	–	75.0	–	–

18

2:4-Dinitro Chlorobenzene (2:4-DNCB)

Preparation of 88% Mixed Acid

In S/S/ Vessel

Charge

860 Kgs 98% — 814 Kgs Real HNO_3

Cool to 10°C using CHW in jacket

Add in 30 mts.

33 Kgs 98% — H_2SO_4 –Temp will rise upto 20°C

Add in one hour

61 Kgs 60% — Oleum-Temp will rise upto 30°C

Volume will be about 600 litres

= 925 Kgs

stir 1/2 at 25-30°C and

Analyse

88.0% W/wt. HNO_2

11.5% H_2SO_4

0.8% $NaHSO_4$

0.5% Water

Use for the Nitration Step-I—35% Mixed Acid

Charge

910 Kgs 98% — 892 Kgs Real – HNO_3

Cool to 10°C

Slowly in 3 to 4 hrs.

Add

1310 Kgs 98%	H_2SO_4 at 10 to 25°C using
	chilled water in jacket
	Then cool to 10°C again
	Add
329 Kgs 60%	Oleum in about 4 hrs.
	Temp from 10°-25°C
	Stir for one hr. at 20-25°C
	and Analyse
	Volume 1600 litres = 2550 Kgs
	35% HNO_3
	65% H_2SO_4
	No water
	0.8% $NaHSO_4$

Use for Nitration Step-II

	Conical Vessel of 4000 litres capacity
	Charge
1672 Kgs 98%	H_2SO_4
	In one hour-Add
428 Kgs	Ice-Stir for 30 mts.
	Analyse the strength
	It should be 77.5 to 78.5%w/wt
	Charge
1500 Kgs	MCB Fresh
	Stir 15 mts and
	Add in 5 to 7 hrs.
	88% Mixed Acid
	at 45-50°C using Chilled Water
	in jacket for controlling the temp.
	volume will be-3200 litres
	Heat to 70°C
	Stir for 4 hrs at 70°C
	Check S. P. of Oily layer
	It should be 49.5 to 51.0°C
	Oily layer is a mixture of

33% OCNB and 67% PCNB
Cool to 60°C and
Settle for 2 hrs and
Drain bottom acidic layer
of 70% Spend Acid
to MSLL Storage tank use
this spent acid for the next
batch after adjusting strength
to 78% Volume of Oily layer
1600 to 1650 litres.
Then slowly in 7 to 9 hrs.
Add 35% Mixed Acid
to the Oil at 50 to 55°C temp.
using chilled water in jacket for
controlling the temp.
Heat to 80°C
Stir at 80°C for 5 hrs.
Analyse oily layer for
PCNB content
It should not be more than 1%
If more-Stir for 2 hrs more
At 80°C and Re-analyse
If okay
Cool to 60°C
Settle for 2 hrs.
Drain acidic layer to gutter
Volume of oil
1600 to 1650 litres
Add 2000 litres water of 60°C
Stir for one hour at 60°C
Settle and Drain water layer
to gutter using Short blow leg of the Vessel
Again add 2000 litres water of 60°C
Stir and add about 8 to 10 Kgs

Soda ash and make alkaline to
PH 7.0 to 7.5
Settle for one hour and
Drain alkaline layer to gutter
Using short blow leg.
Give one more simple water wash
Of 2000 litres at 60°C-Stir one hour
Settle for one hour
and Drain water wash using
shortblow leg of the vessel
Take Vacuum in Vessel
(350 to 400 m. m)
Heat to 80 to 90°C and
Dehydrate the oil layer
Cool to 60-65°C
and fill in drums. Wt. of 2:4 DNCB-2650 Kgs.

Equipments

1. Storage Tank of S/S of 20000 litres for storing 98% HNO_3
2. Storage Tank of S/S of 2000 litres for storing 35% Mixed Acid.
3. Storage Tank of S/S of 1000 litres for storing 88% Mixed Acid
4. M/S Storage Tank of 12500 litres for 98% H_2SO_4
5. M/S Storage Tank of 12500 litres for 60% Oleum
6. MSLL Storage Tank for 70% Spent Acid (H_2SO_4)
7. 500 litres S/S vessel for Preparing 88% Mixed Acid and 35% Mixed Acid
8. 4000 litres S/S Conical Vessel for process with Short Blow leg at 1750 to 1800 litres for removing washes and Bottom outlet for draining acid layer and for filling 2:4 DNCB
9. S/S Measure Vessel of 1000 litrs capacity for 88% Mixed Acid
10. S/S Measure vessel of 200 litres capacity for 35% Mixed acid
11. M/S Storage Tank of 2000 litres for storing MCB

19

Manufacture of Sulfo-J-Acid (2-Amino-5-Naphthol-1:7-Disulfonic Acid)

1. Principle:

J-Acid (2-Amino-5-Naphthol-7-Sulfonic Acid) is Sulfonated with 65% Oleum to give 2-Amino-5-Naphthol-1:6:7-Trisulfonic Acid. The sulfonation mass is then diluted with Sulfuric Acid and drowned into ice-water mixture. It is then heated to hydrolyse the Sulfonic acid group in position 6. The resulting Sulfo-J-Acid liquor is then clarified and the objective product isolate from the clear filtrate by salting. The wet product is evaluated for percentage Sulfo-J-Acid content and is used or sold as such in wet form.

2. Reactions:

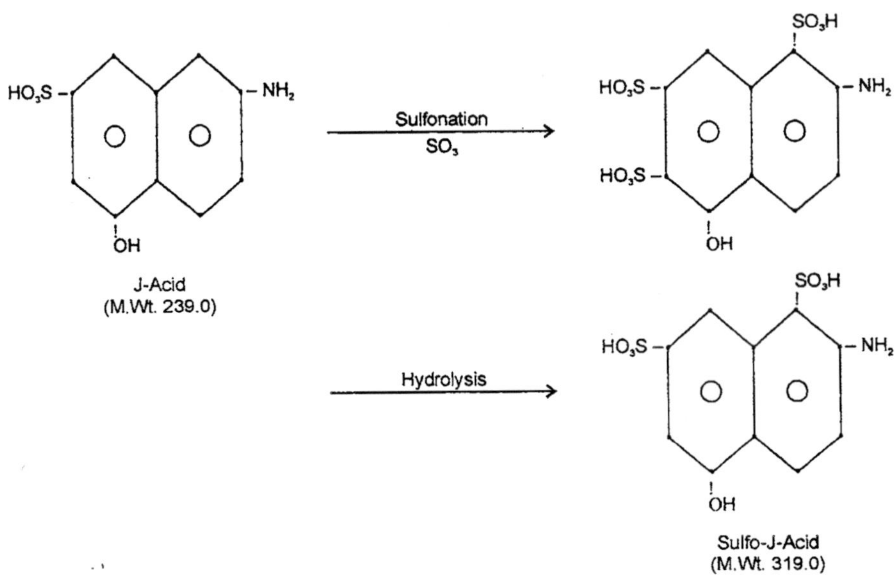

3. Process-Procedure:

3.1. Sulfonation

Tool : A 3,000 L capacity C. I. jacketed Kettle equipped with an Anchor-type, C. I. Agitator driven by a 5 H. P., 3-phase, closed Electric motor connected with a Reduction-gear to give 50 r. p. m. speed to the agitator. The Kettle is further equipped with a Temperture-Recorder-cum-Control, a Pressure-gauge, a Venting system connected to a Scrubber, a manhole, an acid-charging line and a blow-out line lading to the Drowning Vessel.

Check tool.

The Sulfonator must be clean and dry.

Measure

2800.0 Kgs of as is 65% Oleum corresponding to

1820.0 Kgs of Real Sulfur Trioxide (M. Wt. 80) in a suitable Measure Vessel and run it into the clean and dry Sulfonation Kettle described above.

Start agitator.

Circulate chilled water in the Sufonator jacket and cool the Oleum to 12°C.

This takes about 2-1/4 hours.

Weigh

680.0 Kgs of Real, Dry J-Acid powder (M. Wt. 239) having moisture content not exceeding 4%.

Charge the J-Acid powder, evenly over a period of 5 hours into the Oleum maintaining the Sulfonation temperature between 40-45°C by controlling the chilled water flow in the jacket.

Raise the temperature of Sulfonation to 90°C slowly over a period of 1-1/4 hours by regulating a controlled passage of steam in the jacket.

Stir the sulfonation for 2 hours at 90-92°C. Stop heating.

Drain down steam in the jacket.

Cool the batch to 50°C during about 2 hours

by circulating ordinary water in the jacket.

The sulfonation is positively complete. There is no test for checking the completion of sulfonation.

Then gradually charge in the batch

700.0 Kgs as is 80% Sulfuric Acid over about 1-1/2 hours

allowing the temperature of the sulfonated mass not to exceed 80°C.

Stir the batch at 80°C for 30 minutes.

Measure volume of the batch; it will be about

2,100 L at 80°C.

3.2. Drowning & Hydrolysis

Tool : A 15,000 L capacity, M. S. R. L. Tile-lined Kettle equipped with a gate-type or paddle-type Silicon-iron Agitator driven by a 7.5 H. P., 3-Phase, closed electric motor connected with a Reduction-gear to give 60/40 r. p. m. speeds to the agitator. The Kettle is further equipped with a Temperature-Recorder cum Control, a Pressure-gauge, a Venting-system leading to the Scrubber, a direct steam line, an immersion cooling coil of lead attached with a Brine-cooling arrangement, a manhole and bottom-discharge line leading to the Screening Press.

Check tool.

In the clean Drowning Kettle,

Charge the Brine wash liquor, preserved from the previous batch and adjust its volume to

7,000 L at 8-10°C by adding requisite quantity of crushed ice and treated water.

Set the agitation at fast speed of 60 r. p. m.

Start circulation of chilled Brine in the cooling coils. Then slowly blow the Sulfonated mass into the Drowning Kettle at such a rate that the temperature of the drowning mixture does no rise above 15°C. The flow of chilled Brine in the cooling coils should be kept continuous, throughout the drowning.

Note : The temperature during drowning is strictly

Maintained below 15°C so as to prevent the back hydrolysis of Sulfo-J-Acid to J-Acid.

The drowning of the sulfonated batch takes about 5 hours. Measure volume of the drowned batch and record.

Adjust volume of the batch to

10,000 L with treated water.

Stir the batch for 1/2 hour without temp. control.

Measure volume and temperature and record.

Draw a sample of the drowned batch and determine

Total amines as a conversion to Sulfo-J-Acid of

Theory in the drowned batch. Including the product coming from the brine-wash of previous batch, the total conversion will be about 97-98% of Theory.

Determination of conversion as Sulfo-J-Acid

Measure 200 ml of the well-stirred sample of the batch suspension and filter it through a buchner funnel. Wash the cake in the funnel with 600 ml of cold water. Transfer the filtrate quantitatively into 1 L Volumetric flask and dilute to the mark with more of water. Mix the volumetric solution well. Pipette a suitable aliquot of this solution into a litre beaker and titrate at 15°C with a Standard N/2 Sodium Nitrite solution to a 3-minute end point on KI-Starch papers.

The,

$$\frac{\text{ml. titre} \times \text{N. factor of Nitrite soln.} \times \text{V}}{\text{ml. Aliquot taken} \times 6}$$

= % Conversion of Theory as Sulfo-J-Acid (M. Wt. 319.0) where,

V = Volume of the batch in litres.

Then slowly heat the batch to 70°C over a period of 2 hours with coil steam.

Stir the hydrolysis strictly at 70-72°C for 4 hours.

The hydrolysis of the Sulfonic acid group in position 6 is positively complete by now. There is no test for checking the completion of Hydrolysis.

Then add to the batch

70.0 Kgs of Active Carbon and

125.0 Kgs of Hyflosupercel.

Stir the batch at 70-72°C for 2 hours more.

The batch is now ready for screening.

Measure final volume of the batch; it will be about

10,250 L at 70-72°C.

3.3 Screening the Batch

Set up a Filter Press clothed with Terylene filter cloths. Then pump the hydrolysed batch held at 70°C as quickly as possible through the press, the clear filtrates running into the Isolation Vessel described under Step 3.4.

During screening, the temperature of the batch may fall to 60°C. The clarification takes about 4-1/2 hours. When the batch is filtered through, charge

1,000 L of treated water in the Drowning Kettle, steam it to 40°C with direct steam and pass through the Press during about 1-1/2 hours to wash down the occluded product in the carbon-filter aid cakes.

Run this washing into the Isolation Vessel holding the main batch filtrate. Washing takes about 2 hours.

Then air the Press to no drip.

Dump the Press and discharge the wet cakes in tarred casks. Weigh and record the weight of this sludge.

Determine the water-soluble Amine and total Amine

Contents of the well mixed wet sludge sample as described below.

Weight of Carbon-Hyflo wet sludge,

= Approx. 450-500 Kgs containing

about 200 Kgs of dry solids.

This wet sludge is finally washed off to drains.

Analysis of Carbon-Hylfo wet sludge.

(a) Determination of Water-Soluble Amines

Weigh 50 gms of the well mixed wet sludge sample and transfer to a 1 L beaker with about 500 ml of water. Stir the sample to a slurry and warm to 50°C. Filter the slurry on Buchner funnel.

Wash the cakes with 100 ml of more water and suck well. Transfer the filtrate and washing to a clean 1 L beaker and add 20 ml of Conc. HCl. Add ice to cool to 15°C and titrate, under agitation, with a Standard N/2 Sodium Nitrite solution to a 3-minute end point on KI-Starch papers.

Then,

$$\frac{\text{ml. Burette titre} \times \text{N. factor of Nitrite sol.} \times \text{Wr}}{\text{Ws} \times 30}$$

= % Loss at water soluble Amines of Theory as Sulfo-J-Acid (M. Wt. 319).

Where,

Wr = Weight of total wet sludge in Kgs.

Ws = Weight of Sample taken in Gms.

(b) Determination of Total Amines

Weigh 25 gms of the well mixed wet sludge sample and transfer it to a 1 L beaker with 500 ml of water. Stir to a slurry and add to it 40% Caustic Soda solution to obtain a distinct faint-red test on Phenolphthalein paper. Warm the solution to 50°C and confirm the test. Then add 20 ml of Conc. HCl (or more to obtain a strong-blue Congo Red test) and enough ice to cool the solution to 15°C. Finally, titrate the solution under agitation with a Standard N/2 Sodium Nitrite solution to a 3-minute and point on KI-Starch papers.

Then,

$$\frac{\text{ml. Burette titre} \times \text{N. factor of Nitrite soln.} \times \text{Wr}}{\text{Ws} \times 30}$$

= % Loss as total Amines of theory as

Sulfo –J-Acid (M. Wt. 319).

Wr = Weight of total wet sludge in Kgs.

Ws = Weight of sample taken in gms.

3.4 Isolation and Filtration

Tool : A 15,000 L capacity M. S. R. L Tile-lined Kettle equipped with a paddle-type Wooden or M. S. Lead-lined Agitator driven by a 10 H. P., 3-phase, closed Electric motor connected with a Reduction-gear to give 30 r. p. m. speed to the agitator. The Kettle is further equipped with a Thermometer having thermo-well a lead, a Vent-line a Pressure-gauge, a manhole, an immersion Cooling coil of Lead attached with Brine-cooling arrangement, a direct steam line and a bottom-discharge line leading to the Filter Press.

This Vessel is holding the screened batch liquor from the previous step.

Start agitator in the Isolation Vessel.

Measure temperature of the screened batch liquor and washings and adjust it to 50°C.

Add to it

10.0 Kgs of Lissapol C paste and stir to solution.

Draw a sample of this screened liquor for determination of total Liquor-yield and the Salt-charge.

Measure volume of the batch and adjust it to

11,500 L at 50°C.

Determination of total Liquor-yield and Salt-Charge

(a) Yield determination

Pipette 50 ml of the screened liquor sample at 50°C and place it in a 500 ml Volumetric flask. Dilute to the mark with more of water and shake well.

Pipette a suitable aliquot of the Volumetric solution in a 1 L Beaker. Add 500 ml of water and enough ice to cool to 15°C. Titrate it, under agitation, with a Standard N/2 Sodium nitrite solution to a 3-minute end point on KI-Starch paper.

Then,

$$\frac{\text{ml. Burette titre} \times \text{N. factor of Nitrite Soln.} \times \text{V}}{\text{ml. aliquote of volumetric soln.} \times 3}$$

= % Total Liquor yield as Sulfo J-Acid of Theory (M. Wt. 319)

Where,

V = Volume of total screened liquor in litres.

N. B.: This total Liquor yield normally amounts to about 80% of Theory at this stage.

(b) Determination of Salt-Charge

Calculate the quantity of Common Salt charge at a rate of 20% on the batch volume.

The volume of the clear screened batch liquor is 11,500 L.

Therefore, the charge of Common Salt necessary for isolation.

= 2,300 Kgs.

To the screened liquor in the Isolation Vessel maintained at 50 °C charge a total of

2300.0 Kgs of Common Salt in such a way that 1/5th of the Salt quantity is charged evenly during first half hour.

Stir for 1 hour at 50 °C.

Then charge the remainder 4/5th quantity of Salt during 3 hours, the temperature of Isolation being maintained at 48-50 °C, throughout.

Stir the Isolation of 1 hour more a 48-50 °C.

Then start circulation of chilled water in the cooling coils and cool the batch slowly to 20 °C over a period of 6 hours.

Stir the batch further for 2 hours at 20 °C.

Measure volume of the Isolation mass at 20 °C and record.

Draw a sample of Isolation and determine the liquor-loss as described below.

Determination of Liquor-loss

Measure 500 ml sample of the well-stirred isolation slurry in Measuring Cylinder and filter it through Buchner funnel. Suck the cake in the funnel well but do not wash the cake with water.

Measure volume of the clear filtrate and record (Vs).

Transfer 200 ml of the clear filtrate into a 1 L beaker. Add 400 ml of cold water and enough ice to cool to 15 °C. Then add 25 ml of 30% HCl and titrate under agitation with a Standard N/2 Sodium Nitrite solution to a 3-minute end-point on KI-Starch papers.

Then,

$$\frac{\text{ml. Burette titre} \times \text{Norm. of } NaNO_2 \text{ Soln.} \times \text{Vs} \times \text{Vi}}{907.6 \times 1000}$$

= % Liquor-loss as Sulfo-J-Acid on Theory (M. Wt. 319.0) Where,

Vs = Volume of sample filtrate in ml.

Vi = Total Volume of Isolation mass in litres.

Now, if the liquor-loss is greater than 20% of Theory. Stir the Isolation mass further at 20 °C and test at 3-hour intervals until the difference between two liquor-loss determinations is less than 0.3% of Theory.

Finally, pump the isolated batch at 20°C to Filter Press clothed with Terylene Filter cloths.

Release the filtrates to acid-drains. Suck the Press well to drip drown the M. L. to a maximum.

This takes about 2-1/2 hours.

Then charge in the empty Isolation Kettle

1800 L of 30%m w/vol Brine (540 Kgs Salt) and pump it through the Filter Press slowly during about 45 minutes to serve a mild wash on the press cakes.

Collect these Brine washings and preserve them for

Use in the Drowning operation of the subsequent batch. Sucky dry the cakes by blowing air.

Dump the Press and collect the wet product in suitable containers.

Weigh the wet cakes and record.

Prepare a representative sample from the batch wet cakes. Determine real Sulfo-J-Acid on M. Wt. 319.0.and record.

4. Yield:

Weight of Wet Cakes,

= Approx. 1300-1400 Kgs containing about

36-38% of real Sulfo-J-Acid.

= 500 Kgs of Peal Sulfo-J-Acid (M. Wt. 319).

= 55.% of Theory on real J-Acid charge.

4.1. Specification of Sulfo-J-Acid

The maximum allowable limit of J-Acid in Sulfo-J-Acid

is 4% on 100% basis.

5. Raw Materials Usage Data:

	Names of Raw Materials	*M. Wt.*	*Quantity Real (Kgs)*	*Moles.*	*One Mole Ratio*
i)	J-Acid, dry powder	239.0	680.0	2.8452	1.0000
ii)	Oleum, 65%	–	2,800.0	–	–
	SO_3 :	80.0	1,820.0	22.7500	7.9959
iii)	H_2SO_4, 80% - as real	98.0	560.0	5.7143	2.0084
iv)	Activated Carbon	–	70.0	–	–
v)	Hyflosupercel	–	125.0	–	–
vi)	Lissapol C, paste	–	10.0	–	–
vii)	Common Salt (Isolation)	–	2,300.0	–	–
viii)	Common Salt (Brine-wash)	–	540.0	–	–

20

Manufacture of N-Methyl-J-Acid

1. Principle:

J-Acid is condensed with Monomethylamine in presence of Sodium Bisulphite (Buchrer Reaction). The condensation is carried out under reflux for about 12 hours and excess Methylamine and Ammonia distilled out. Finally, the condensed mass is drowned into Sulfuric Acid Solution, the Sulfur Dioxide gas removed by aeration and the product isolated by filtration. The wet, crude N-Methyl-J-Acid is finally purified. The purified wet product is evaluated for real content and is used or sold as such, in wet form.

2. Reactions:

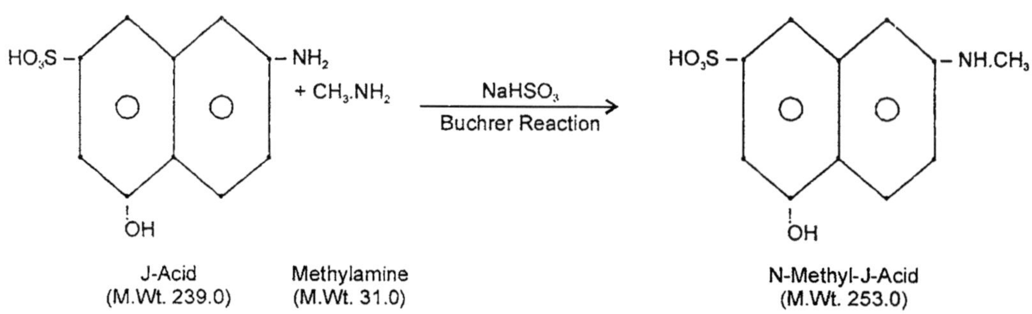

3. Process-Procedure:

3.1. Condensation

Tool A 3,000 L Capacity SS, Jacketted Kettle equipped with an anchor-type SS or M. S. R. L. Agitator driven by a 5 H. P., 3-phase, closed Electric motor connected with a Reduction-gear to give 40 r. p. m. speed to the agitator. The Kettle is further equipped with a Temperature-Recorder-cum-Control, a Pressure-gauge, a Reflux-cum-Distillation arrangement with an M. S. Distillate-Receiver, a manhole, a Lead or SS immersion coil and a blow-out line Leading to the Isolation Kettle.

Check tool.

In the clean Condensation Kettle, charge

775.0 L of treated water. Start agitator and charge

675.0 Kgs of real Sodium Bisulfite (M. Wt. 104.0). Stir for 15 minutes. Measure volume of the charge and record. Adjust the volume to

1,250.0 L with more of treated water. Stir the Bisulfite to solution. The resulting solution contains about 40% w/w sodium Bisulfite. Check the strength of the solution analytically and if necessary, adjust it to 40%w/w sodium Bisulfite content. Then charge into the Bisulfite solution.

621.0 Kgs of real J-Acid (M. Wt. 239.0) as wet cakes of 50%. Stir for 20 minutes and charge in 2 to 3 hours, maintaining the temperature to 50-55°C

322.0 Kgs of real Monomethylamine (M. Wt. 31.0) as 40% solution. As a result of Methylamine addition, the temperature of the charge rises to about 5-55°C. Set the Condensation Kettle for reflux. Stir the batch at 50-55°C for 30 minutes. Check the pH; it will be 9.3-9.7. Then add to the batch, evenly during about 45 minutes, a fixed charge of

100.0 Kgs of as is 98% Sulfuric Acid corresponding to

98.0 Kgs of real Sulfuric Acid. The temperature of the batch rises to about 60-62°C as a result of the acid addition. Volume of the batch will be about

3000.0 L at 60-62°C. If not, adjust it to this limit. Then heat the batch slowly to the reflux temperature of 105°C over about 2 hours by passing steam in the Kettle Jacket. While the batch is being heated to the reflux temperature of 105°C, draw a sample of the reaction mass at 75°C and note its pH. The pH should be about 7.5. Do not adjust the pH. The pH should be about 7.5. Do not adjust the pH, but simply record it. Finally, heat the batch further to the reflux temperature of 105°C. Again draw a sample of the reaction mass at 105°C and not its pH; it will be about 7.5. Do not adjust the pH, but simply record it. Reflux the batch at 104-105°C for 12 hours.

Note : Do not reflux at a temperature exceeding 106°C since frothing is likely to take place at that temperature.

At the end of 12 hours reflux, note pH of a batch sample. The pH will be about 7.4. Again do not adjust the pH, but simply record. Close the reflux line and set the Vessel for distillation. Start distillation of he batch at 105-106°C and distil aqueous Ammonia CH_3NH_2 mixture at a rate of 65 litres per hour until a total volume of 725 litres is distilled over. At the end of distillation, note pH of a batch sample; it will be about 6.7-6.8. Do not adjust the pH. Distillation period should not be longer than 14 hours maximum. Prolonged distillation over 14 hours gives lower yield. If necessary, circulate a passage of steam in the Immersion Coil to achieve necessary distillation rate.

N. B. : If in case, 14 hours of distillation period fails to produce 725 litres of distillate, measure pH of a batch sample. If the pH is not lowered to 6.7-6.8, distil the batch further for some time until the said pH is achieved or the objective distillate volume is distilled over, whichever is earlier.

In fact, it is strongly suggested that the distillation should proceed a sufficiently faster so as to distil over the said distil over the said distillate volume during about 11 to 12 hours.

Near the end of distillation, the batch temperature rises to about 107-108°C. At the end of distillation, shut off steam in the Coil and jacket. Dilute the batch by adding.

300.0 L of treated water and cool the batch to 85°C by circulating ordinary water in the Kettle jacket.

Note : The condensed mass is not stable and as such, it is suggested that isolation should proceed without any delay. Prolonged delay in isolation causes appreciable yield losses.

Measure final volume of the condensed batch and adjust it to

2,500.0 L at 85°C.

Blow the batch to Isolation Vessel in 4 to 5 hours, maintaining the temperature to 85 ± 2°C.

3.2. Isolation

Tool : A 6,000 L capacity M. S. Lead-lined jacketted Kettle equipped with a gate-type, M. S. Lead-lined Agitator driven by a 5 H. P., 3-Phase, closed Electric motor connected with a Reduction-gear to give 40 r. p. m. speed to the agitator. The Kettle is further equipped with a Thermometer having lead Thermo-well, a Pressure-gauge, a Venting-system attached to a scrubber for absorbing SO_2 gas, a lead line reaching the bottom of the Kettle for blowing compressed air for SO_2 removal, an Immersion Cooling Coil of lead for chilled water circulation, a manhole and a bottom-discharge line leading to the Filter Press.

Check tool.

In the clean Isolation Kettle, charge

1,100.0 L of treated water. Start agitator and charge

725.0 Kgs of as is 98% Sulfuric Acid corresponding to

710.5 Kgs of real Sulfuric Acid. Note the temperature of the charge and heat to 85°C with jacket steam.

Add

3.0 kgs of Calsolene Oil HS. Vent the Isolation kettle to Sulfur Dioxide absorption system. Then receive in the condensed batch mass from the Condensation Vessel at such a rate that the evolving Sulfur Dioxide gas does not overload the absorption system. Maintain the temperature, during drowning, in the range of 84-87°C by circulating chilled water in the Cooling Coil. While about 3000 L of the condensation mass still remains to be drowned, stop further receipt and allow the drowned mass to crystalise in a satisfactory physical form. When the crystallization appears to be satisfactory, receive in the remaining 300 L mass.

Maintain the isolation temperature at 84-87°C, throughout the drowning operation. Take care to avoid excessive frothing. Wash down the inside wall of the empty condensation Vessel with

300.0 L of treated water. Agitate this wash and heat to 85°C. Transfer the wash to the Isolation Vessel holding the drowned batch. On completion of drowning, blow a passage of air into the isolation mass and continue bubbling air for 4 hours at 85-87°C so as to completely drive off suffer Dioxide gas. During the SO_2 boil-off, the isolation mass must show a strong blue test on Congo Red papers; if not, add corrective quantity of 98% Sulfuric Acid to keep the test positive, throughout. Finally, cool the batch slowly to 20-23°C during 5 hours by circulating chilled water in the cooling coil. Measure volume of the batch; it will be about

4,300-4,500 L at 20-23°C.

Draw a sample of the product slurry and determine liquor-loss in the M. L. as follows:

Determination of Liquor-loss

Measure 500 ml of the well-stirred isolated batch slurry in a 500 ml Measuring Cylinder. Filter it through a dry Boucher funnel. Suck the cake well but do not wash. Measure volume of clear filtrate and record. Pipette 50 ml of the clear filtrate in a 1 L beaker. Add 500 ml of ice-cold water and stir mechanically. Then add to it 5% Caustic Soda solution, dropwise, until the diluted filtrate shows no blue test on Congo Red papers. Finally, add to it 100 ml of 20% Soda Ash solution and cool to 3-5°C by adding ice as required.

Titrate the mixture with a Standard N/10

p-Chloro-Aniline Diazo solution until a spot of the Coupling mixture on a piece of filter paper when cross-spotted with a drop of Alkaline-H-Acid solution produces a pink line at the junction of the spots.

Let,

V= Batch volume in litres.

V= Volume of 500 ml batch sample filtrate in ml.

T= Burette titre in ml.

Then,

$$\frac{T \times N.\text{ factor of p-CA Diazo} \times v \times V \times 154}{10^8}$$

= % Liqur-loss of Theory.

and

$$= \frac{T \times N.\text{ factor of p-CA Diazo} \times v \times V \times 1012}{10^8}$$

= Kgs liquor-loss in total batch Mother liquor.

Note : The liquor-loss should not be greater than 110 Kgs in the total batch Mother liquor which corresponds to about 16.7% of Theory.

If the liquor-loss from the above determination is found to be higher than the given limit, stir the batch further for about 3 hours at 20-23°C and recheck the test. The batch slurry, however, must be strongly Congo-Acid, throughout.

If the liquor-loss is still high, no further action is taken and the batch at 20-23°C through a Filter Press. Suck the M. L. Well.

Charge in the empty Isolation Vessel, water and enough ice to make up

400.0 L of wash-water at 15°C.

Stir for 5 minutes at 15°C and pump this wash through the Press giving a mild wash on the Press cakes. Discard the M. L. and washings to acid drains. Suck the press cakes to no drip. Dump the Press. Centrifuge the Press cakes. Weigh the centrifuged cakes and record. Prepare a well-mixed representative sample from the centrifuged cakes and estimate real N-Methyl-J-Acid content on M. Wt. 253.0 by taking a Coupling titre with a Standard N/10 *p*-Chloro-Aniline Diazo solution. Calculate yield and record.

4. Yield:

Weight of centrifuged cakes,

= Approx. 1000-1100 Kgs

= 538 Kgs average Real N-Methyl-J-Acid (M. Wt. 253)

= 82% of Theory on real J-Acid charge.

5. Raw Materials Usage Data:

	Names of Raw materials	*M. W.*	*Quantity Real (Kgs)*	*Moles*	*One Mole Ratio*
1.	J-Acid	239.0	621.0	2.5983	1.0000
2.	Sodium Bisulfite	104.0	675.0	6.4904	2.4979
3.	Monomethylamine	31.0	322.0	10.3871	3.9976
4.	Sulfuric Acid, 98% (Fixed)	98.0	98.0	1.000	0.3848
5.	Sulfuric Acid, 98% (Isolation)	98.0	710.0	7.2500	2.7903
6.	Calsolene Oil HS	–	3.0	–	–

Purification of Crude N-Methyl-J-Acid

1. Process-Procedure:

1.1 Dissolution of Crude

Tool : A, 3000 L capacity M. S. R. L. Kettle equipped with a gate-type M. S. R. L. Agitator driven by a 3 H. P., 3-phase, closed Electric motor connected with a Reduction-gear to give 30 r. p. m. speed to the agitator. The Kettle is further equipped with a Thermometer, a vertical vent, a Pressure-gauge, a manhole, an SS or Lead immersion cooling Coil and a bottom-discharge line leading to the Filter Press.

Check tool.

In the clean Kettle,

Charge

700.0 L of treated water. Start agitator and charge

1000-1100 Kgs centrifuged wet cakes of Crude N-Methyl-J-Acid containing

538.0 Kgs of Real, N-Methyl-J-Acid (M. Wt. 253.0). Stir to a smooth slurry. Displace air in the Kettle by blanketting the upper surface of the charge with an inert gas (Nitrogen.)

Note : Maintain the blanketing of Nitrogen gas until precipitation with Formic Acid is over and the batch has been cooled down to 20-22°C.

Now to the smooth slurry, add gradually in about 3 hours with frequent testing, 40%w/w Caustic Soda lye until the solution imparts a red test on B. Y. papers. Note pH; it must be in the range of 7.8-8.0. During Caustic Soda addition, the temperature of the batch rises to about 40°C. Adjust the temperature at 40°C by passing steam in the immersion coil, if necessary. Confirm positive red test on B. Y. papers and adjust the pH in the said range.

Draw a sample of the solution in a 1-L beaker and observe for any undissolved product still remaining in suspension. Stir the solution at 40°C for 20 minutes under the blanket of the Nitrogen gas.

Confirm the product to be in complete solution

Measure volume of the solution and adjust it to

1600-1800 L at 40°C and stir or half hour at 40°C under the blanket of Nitrogen. Note and confirm the pH to be in the said range. Record the quantity of Real Caustic Soda required for dissolution. About 90-95 Kgs of real Soda is required. The solution is not clarified and is subjected to precipitation, as such.

1.2. Precipitation and Filtration

Then run into the solution a charge of

250.0 Kgs of 30% w/w HCl, corresponding to

75.0 Kgs of real HCl, evenly over a period of 3-4 hours maintaining the temperature at 40°C.

Note pH; it should be between 1.5-2.0

If not, adjust in the said range with HCl.

Measure volume; it will be about

2300-2400 L at 40°C.

Stir for 30 minutes

The product precipitates out in the fine form.

Confirm the pH.

Cool to 18-20°C by circulating cooling water and chilled water in the coil in 3-4 hours.

Check the liquor loss; I should not be more than 7%.

If the liquor loss is more, stir the batch 18-20°C for 2-3 hours more and check the loss again.

Filter the bach through Filter-Press.

Suck the M. L.

Dump the Press and Centrifuge the wet Cakes.

2. Recovery

Weight of Centrifuged wet cakes,

= Approx 700 Kgs of 72% C. Value as N-Methyl-J-Acid

= Approx 490-500 Kgs of real N-Methyl-J-Acid (M. Wt. 253.0)

= 93% Recovery on the charge of crude material.

21

Manufacture of Meta Phenylene Diamine (MPD)

1. Procedure:

		Charge
		Some T. Water and
500	kgs	Iron broings. Add
55	Kgs	as is 30%w/w., HCl. Steam to boil during 10-15 minutes.
		Run in
280	Kgs	Real *m*-Dinitrobenzene (molten) during 80-110 minutes. Run in an equal steam of cold water along with the molten Dinitrobenzene to maintain proper reduction rate. If the temp is below 100°C. Stop Meta-Dinitrobenzene addition and heat to 100°C.

A spot on filter paper should never show any trace of yellowness or oil in the soluble iron when spotted with 5% Na2S solution.

The reduction-mixture should always show a positive pink test on Blue litmus. If not. HCl in 4 Kgs lots should be added till the test is obtained. If soluble iron is not there, iron borings should also be added to the extent of 40-50 Kgs.

Cold water is added to avoid boil over.

The reduction mass should never boil over.

When MDB addition is over, stop cold water addition.

Stir at boil until reduction is complete. When no yellowness and no oil is seen when spotted on a filter paper, the reduction is complete.

Sir at boil for 15 minutes after reduction is complete.

Then add.

25	Kgs	Real Soda Ash as a 20% solution to a slight test on B. Y. Add more Soda Ash if necessary. No soluble iron should be seen when spotted with 5% Na2S solution. Stir for 15 minutes holding slight test on B. Y.

Add,

6 Kgs Kieselguhr. Stop stirrer and filter.

2. Raw Materials Usage Data:

	Name of Raw Material.	*M. Wt.*	*Real*	*Moles*	*One Mole Basis*
1.	*m*-Dinitrobenzene	168.0	280	1.670	1.00
2.	Iron borings	56.0	500	8.960	
3.	30% HCl w/wt.	36.5	16.5	0.452	
4.	Soda Ash	106.0	25.0	0.25	
5.	Kieselguhr	–	6.0	–	

3. Yield:

MPD Approx. 2600 Kgs. 6-7%

Solu. Containing 155.4 Kgs.

Real MPD = 86.3% Theory.

22

Manufacture of N-Phenyl-Peri Acid (1-Phenylamino-Naphthalene-8-Sulfonic Acid)

1. Principle:

Peri Acid (1-Naphthylamine-8-Sulfonic Acid) is condensed with excess of Aniline Oil. Much of the unreacted excess of Aniline Oil is recovered by vacuum distillation. The reaction mass is then basified and steam-distilled to strip off the remaining Aniline. The sodium salt solution of the final product is treated with activated carbon, clarified and from the clear filtrate, free N-Phenyl-Peri Acid is precipitated with mineral acid.

2. Reactions:

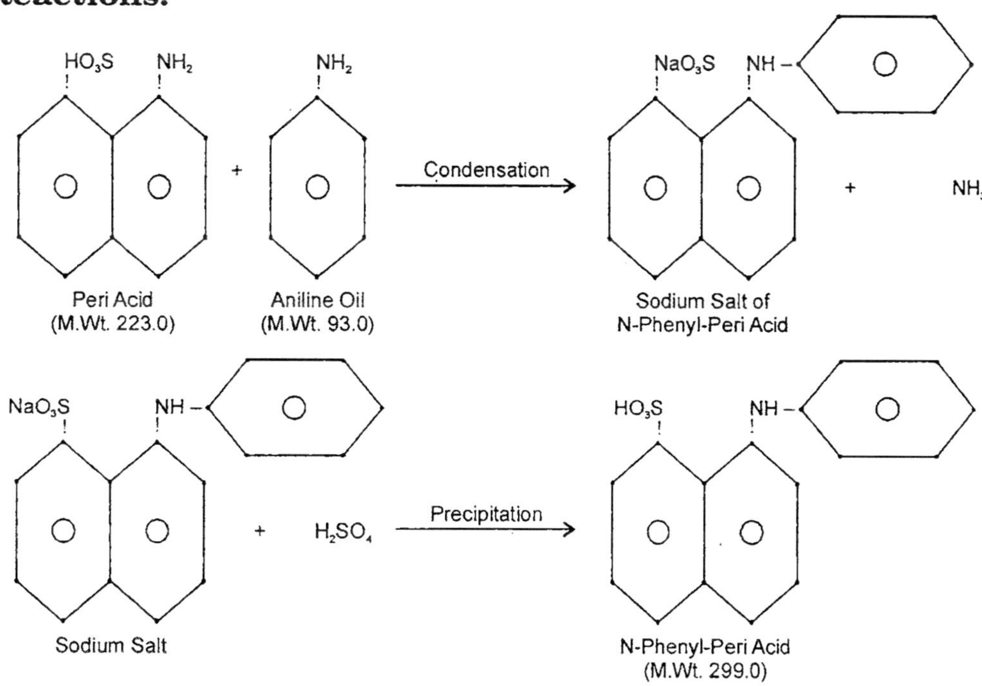

3. Process-Procedure:

3.1. Condensation

Tool A 2000 L capacity Jacketted, SS Kettle equipped with an Anchor-type, SS Agitator driven by a 5 H. P., 3-Phase, closed electric motor connected with a Reduction-gear to give 40/20 r. p.m. speeds to the agitator. The kettle is further equipped with a Temperature-Recorder-cum-Control, a Pressure gauge, a manhole, a jacketed vertical exhaust line for expelling Ammonia, a Vapour line passing through a descent jacketted-Condensor connected with a Condensae Receiver, a safety-valve tested for 2 Atmos. Pressure and a bottom-outlet line leading to the Steam-distillation Kettle.

Check tool.

In the clean, dry Condensation Kettle,

Charge

1,165.0 Kgs of real Aniline Oil (M. wt. 93.0).

Note : Fresh Aniline Oil is used in the first batch of a series. In the subsequent batches, recovered Aniline from previous batches should be used. If the quantity of recovered Aniline is insufficient for a charge, the balance quantity is made up with fresh Aniline.

Start agitator and is at 40 r. p. m. speed.

Then add

50 Grams of Iodine crystals.

Stir for 15 minutes.

Then charge through the manhole

434.0 Kgs of dry, ground, real Peri Acid (M. Wt. 223.0) in quick succession.

The dry Peri Acid powder contains about 95% real Peri Acid; its laurent's Acid content being less than 2%.

The temperature of the charge rises during charging the Peri Acid powder and at the end of charging, it reaches to 70-75°C.

Record the temperature.

The Peri Acid charging takes about 30-40 minutes. Keep the vertical exhaust line open and start circulation of cooling water in its jacket. Also keep the vapour line passing through the Condensor open and circulate cold water through the Condensor jacket.

Close manhole of the Kettle.

Change the speed of agitation to 20 r. p. m.

Start full passage of steam in the jacket of the Kettle and heat the batch to 165-168°C during about 2½ to 3 hours.

Continue the condensation reaction under reflux for 18 hours at 165-168°C.

During condensation, a mixture of poisonous vapours containing Ammonia with some unconensed Aniline vapours is expelled out of the Plant through the vertical exhaust Vent; most of the aniline vapours of course condensing back in the reaction Kettle.

Vapours of some water associated with the reactants of the batch-charge along with some vapours of Aniline distill out during condensation. Mix these condensates with the condensates coming out from the steam-distillation operation of the batch and collectively process them for separation and vacuum distillation to get pure Aniline, as described under Step 3.5.

At the end of 18 hours reflux, draw a batch sample and test as follows for the completion of condensation.

Test : Place 25 gms of the batch sample in a 500 ml beaker, add 100 ml of water and make it faintly alkaline to Phenolphthalein papers by adding 20w/vol Soda Ash solution. Transfer this solution to a 1 L Distillation flask and steam-distill off aniline in the sample.

Cool the contents of the flask to 20°C and filter it through Buchner funnel. The residue of uncondensed Peri Acid on filter paper should not be more than a trace.

If more uncondensed Peri Acid is sighted in the above test continue the condensation of batch further at 165-168°C until the above test is satisfied. Generally, the condensation is complete at the end of 18 hours reflux. See Process Notes.

At the end of completion of condensation, Vacuum-distil the excess of Aniline as follows.

3.2. Vacuum Distillation of excess Aniline

When the condensation is confirmed to be complete, shut of the passage of steam in the Kettle jacket. Drain down steam and water condensates from the jacket.

Cool the batch to 110°C.

Continue stirring at 20 r. p. m. speed.

Connect the Distillation Vent of the Condensation Kettle with a 1000 L capacity, jacketed SS distillate Receiver equipped with a Pressure-gauge, a Thermometer, a blow-out line and Vacuum and.

Connect the Distillate-Receiver with the Vacuum pump.

Now close the Exhaust-Vent on the Condensation Kettle.

Open the valve on the Distillation-Vent leading to the Distillate-Receiver.

When the batch mass in the Condensation Kettle is at 110°C, start vacuum and distil the aniline oil under a pressure of 25-28 mm.

Regulate the vacuum by operating the needle-valve on the vacuum line.

The initial batch temperature of 100°C in the Condensation Kettle will rise to 130°C at the end of distillation of Aniline.

Vacuum Distillation period = 5-7 hours.

Weight of recovered Aniline by Vacuum.

Distillation = About 850.0 Kgs.

Note : This recovered Aniline can be used as such in the subsequent batches.

When the vacuum distillation is over, bring the system to atmospheric pressure.

Change the speed of agitation to 40 r. p. m.

Cool the contents of the Kettle to 95°C.

The charge in the Condensation Kettle

450 L of water followed with

175.0 Kgs of as is 40%w/w Caustic Soda lye corresponding to 70.0 Kgs of real Caustic Soda.

The temperature of the batch falls to about 70°C.

Hold the batch at 70°C until transferred to the Steam-Distillation Kettle.

3.3 Steam-Distillation of Residual Aniline

Tool A 600 L capacity, M. S., jacketed Kettle equipped with a Spade-type Agitator driven by a 5 H. P., 3-phase, closed electric motor connected with a Reduction-gear to give 40 r. p. m. speed to the agitator. The Kettle is further provided with a direct steam line, a Vapour-line passing through two jacketed Condensors and connected with a Distillate-Receiver, a Pressure-gauge, a Temperature-recorder and a bottom-outlet line leading to the Filter Press.

Check tool.

In the clean Kettle

Charge

1,650 L of treated water.

Start agitator and charge

27.5 Kgs of Soda Ash powder and

100.0 Kgs of Calcium Carbonate powder.

Stir for 10 minutes.

Then receive into this the batch mass from the Condensation Kettle in the above step.

Star a passage of steam in the Kettle jacket and heat the batch externally to 90-95 °C.

Star circulation of ordinary water in the jackets of Condensors on the distillation line.

Then inject a passage of direct steam into the batch mass and distil the residual Aniline in the batch.

Regulate the direct steam-flow and also the steam passage in the jacket so as to prevent under rise of the batch volume.

Transfer the distillates from the Distillate-Receiver to a 5,000 L capacity (or larger) M. S., Jacketted Conical Separator described under the step 3.5. of "Processing Steam-distilled Aniline".

Continue steam-distillation till no oily drops of Aniline seem coming out in the distillate.

Determine the Aniline content in separately drawn distillate samples as described in the following test.

Test : Collect separately a fresh sample of the distillate. Pipette 25 ml of this distillate sample in a 500 ml beaker. Add to it 50 ml of water, 20 ml of 30% HCl and enough ice to cool to 0-3 °C. Stir and add to it 1.0 gm of C. P. KBr and finally titrate it with a Standard N/5 Sodium Nitrite solution from a burette to a 2-minute end point on KI-Starch Indicator papers.

Note the burette reading.

If the burette reading is 27.0 ml, the Aniline content of distillate is 2.%.

Collect and preserve the Distillates for separation and recovery of Aniline Oil till the distillates show up to 2.0% Aniline content, as determined by the above test. Do not preserve distillates after the aniline content falls below 2.0%

Time for steam-distillation till this stage = About 6-8 hours.

Now continue steam-distillation of the batch further till the Aniline content of the distillates falls below 0.2%

Discard all these distillates to drains.

Confirm completion of Steam-Distillation with the following test.

Test : Collect separately 250 ml of the distillates sample in 1 L beaker. Add to it 20 ml of 30% HCl and enough ice to cool to

0.3°C. Stir and add to it 1.0 gm. of C. P. KBr and titrate with a Standard N/5 Sodium Nitrite solution from a burette to a 2-minute end point on KI-Starch indicator papers.

Note the burette reading.

If the burette reading is less than 270.0 ml, the Aniline content of distillate has fallen below 0.2%.

When the above test is satisfied, i. e. when the Nitrite titre for a 250 ml distillate sample has fallen below 27.0 ml, the steam distillation is deemed to be complete. The volume of total distillates collected in the Conical Separator for recovery of Aniline is about

4,000-4,500 L

Time taken by this end part of distillation =About 7-8. hours.

Finally, separate and vacuum distil Aniline from the Collected steam-distillates as described under step 3.5. of "Processing Steam-distilled Aniline".

3.4. Isolation of N-Phenyl-Peri Acid

A. Clarification of N-Phenyl-Peri Acid solution:

Measure temperature of the steam-distilled batch in the Steam-distillation Kettle.

Cool the solution to about 85-90°C.

While stirring, charge into it.

78.0 Kgs of Soda Ash powder,

10.0 Kgs of Activated Carbon and

10.0 Kgs of Hyflosupercel (or Kieselguhr).

Stir at 85-90°C for 45 minutes, heating as necessary with jacket steam.

Stop steam in the Kettle jacket.

Drain down steam and water condensates from the Kettle jacket.

Start circulation of ordinary water in the Kettle jacket and stir till the batch temperature falls around 60°C.

Then start circulation of a passage of chilled water in the jacket.

Measure volume of the batch and record.

Adjust the batch volume to

3,800 L at 28-30°C by adding ice and water as necessary and also by adjusting the flow of chilled water in the Kettle jacket.

Then further cool the batch to 20-23°C by making a direct addition of balanced quantity of crushed ice and jacket-cooling, not to allow the batch to exceed

4,000 L at 20-23°C.

The batch is now ready for filtration.

Clarify the batch liquor at 20-23°C

Through a Filter Press and collect the clear batch filtrate in the Precipitation Vessel described under step B.

Wash the Press cakes with

200 L of cold water of 20°C.

Collect the clear washings along with the main batch filtrate held in the Precipitation Vessel.

Suck the Press well to no drip.

Dump the Press.

Discard the Press cakes to drains.

B. Precipitation of N-Phenyl-Peri Acid:

Tool : An 8000 L capacity, M. S. Lead-lined jacketted Kettle equipped with a Spade-type, M. S. L. L. Agitator driven by a 5 H. P., 3-phase, closed electric motor connected with a Reduction-gear to give 60/40 r. p. m. speeds to the agitator.

The Kettle is further equipped with a Thermometer, a Pressure-gauge, an exhaust line, a manhole and blow-out line leading to the Filter Nutsche.

The clear batch filtrate and washings from the above step are collected in this Vessel.

Measure volume of the solution and record;

It will be about

4,200 L at 20-25°C.

Start agitator and stir the solution at 60 r. p. m. speed.

Start circulated of chilled water in the Kettle jacket to maintain the batch temperature at 20-25°C.

Then start adding to it slowly in a broad spray

735.0 Kgs of as is 40%w/w Sulfuric Acid solution corresponding to

294.0 Kgs of real Sulfuric Acid until N-Phenyl-Peri Acid starts falling out as small oil globules.

Foaming takes place during the acid addition.

When precipitation of the product in the form of oil globules is sighted, stop further addition of acid at that stage and continue stirring at 60 r. p. m. speed till the oily droplets begin to solidify.

If necessary, seed the batch at his stage by adding to it about

750 gms of dry N-Phenyl-Peri Acid powder.

The product now starts precipitating out as solid particles.

Now change the speed of agitation to 40 r. p.m. and start the acid addition until the pH of the batch slurry comes in the range of 1.2-1.7.

Stir the batch for 1 hour at 20-25°C and confirm pH of a batch sample in the said range.

If necessary, adjust the pH in the specified range.

The precipitation of N-Phenyl-Peri Acid is complete and the batch is now ready for filtration.

Filter the precipitated product at 20-25°C through an M. S. Brick-lined or M. S. R. L. Nutsche Filter having 7.4 Sq. meter basket area.

Wash the Nutsche cakes with

300 L of cold water of 20-25°C.

Suck the cakes well.

Dump the Nutsche cakes.

Centrifuge the Nutsche cakes.

Dump the centrifuge cakes, weigh and record.

Discard the M.L. and washings to acid-drains. Analyse a representative sample of the centrifuged cakes for N-Phenyl-Peri Acid content.

Calculate and record yield of N-Phenyl-Peri Acid.

3.5. Processing Steam-distilled Aniline

Tool A 5,000 L capacity M. S. jacketed Conical Separator equipped with a Spade-type, M. S. Agitator driven by a 5 H. P., 3-phase, closed electric motor connected with a Reduction-gear to give 40 r. p. m. sped to the agitator.

The Vessel is further equipped with a Vent line, a manhole, a Thermometer and a bottom-discharge line.

The distillates collected from the Steam-distillation operation are stored in this Vessel. Add to this the distillate received from the Condensation Reaction under step 3.1., which also contains some aniline. The volume of these distillates in different batches varies between 4,000-6,000 Litres.

But the average volume of steam-distillates for a normal batch is about

4000-4500 L.

Measure volume of the distillates and record. Star agitator and charge ordinary Common Salt at a rate of

24 Kgs of Salt per every 100 L distillate-volume.

The total quantity of Common Salt required for an average batch will be about

1,100 Kgs. Stir the charge the dissolve the salt.

Stop stirrer and allow to stand to a enable the layers to separate.

Free Aniline oil collects in the upper layer.

Discharge the lower aqueous layer to drains via the bottom-discharge line.

Note : In the event of a larger volume of Steam-distillates, the Separator capacity falls short for a single separation operation. In this case, divide the distillates in two parts to carry out separation.

Draw off the upper Aniline layer and store it in a suitable container for further purification by Vacuum distillation.

From each batch, only about 50-75 Kgs of

Aniline oil is obtained from the separation of Steam-distillates and therefore, the separated oil layers from several batches are stored until a sizable quantity suitable for Vacuum distillation is collected. Finally, vacuum distil the separated oil under a pressure of 25-28 of Hg. Discard the residual sludge in the Distillation still at the end of Vacuum distillation.

Weight the total quantity of recovered Aniline oil and record.

Total Aniline oil recovered from Vacuum Distillation and Steam Distillation operations of one batch,

= Approx. 900.0 Kgs. of real Aniline.

= About 77% of the initial charge.

4. Yield:

Weight of centrifuged cakes,

= Approx, 650-700 Kgs.

= 500 Kgs f real N-Phenyl-Peri Acid (M. Wt. 299.0)

= 86.0% of Theory on the charge of Peri Acid.

Total Aniline Recovery = About 900.0 Kgs.

=77% of the initial charge.

5. Process Notes:

(i) From the experience gained it is observed that the same Aniline can be repeatedly used for at least 10 batches in a series of production campaign without any disadvantage. However, it is recommended that the total recovered Aniline Oil at the end of 10 batches campaign should be subjected to separate Vacuum Distillation before further use. Traces of Ammonia in the regenerated Aniline are not harmful.

(ii) In Leboratory experiments, wet Peri Acid containing about 60% real Peri Acid has been satisfactorily tried out for straight condensation. In this case, abut 5% of the total Aniline charge (i. e. about 58 Kgs of Aniline) is distilled off along with water during Condensation. However, the direct use of wet Peri Acid in the Plan is not tried out.

(iii) If the reaction time is shortened, the condensation remains partially incomplete. If the time is lengthened to 24 hours, there is no adverse effect on the product. If the condensation temperature is raised to 175°C, the yield is lowered by about 5%.

(iv) The solution of Phenyl Peri Acid is clear in dilute alkali. The insolubles consist of traces of Diphenylamine Phenyl Napthalene and other unidentified compounds. These impurities are collectively below 1%.

(v) Aniline oil recovered through Vacuum-distillation in Phenyl-Peri Acid batches is not suitable for use in other purposes. Therefore, it should be used only in the Phenyl-peri Acid manufacture.

6. Raw Materials Usage Data:

	Names of Raw Materials	*M. Wt.*	*Quantity Real (Kgs)*	*Moles*	*One Mole Ratio*
i)	Peri Acid	223.0	434.0	1.9462	1.0000
ii)	Aniline Oil	93.0	1165.0	12.5269	6.4366
iii)	Iodine Crystals	254.0	50 gms.	–	–
iv)	Caustic Soda	40.0	70.0	1.7500	0.8992
v)	Soda Ash	106.0	27.5	0.2594	0.1333
vi)	Calcium Carbonate	100.0	100.0	1.0000	0.5138
vii)	Soda Ash	106.0	78.0	0.7358	0.3781
viii)	Sulfuric Acid	98.0	294.0	3.0000	1.5414
ix)	Common Salt-Ordinary	–	1,100.0	–	–
x)	Activated Carbon	–	10.0	–	–
xi)	Kieselguhr	–	10.0	–	–

23

Manufacture of Peri Acid (1-Naphthylamine-8-Sulfonic Acid)

1. Principle:

Napthalene is sulfonated with 94% Sulfuric Acid and then nitrated with 62% Nitric Acid. The batch is drowned, neutralized with Calcium Cabronate in presence of Magnesium Oxide and the precipitated gypsum filtered off and washed. The product in solution is reduced with Iron and HCl and the precipitated iron-sludge filtered and washed. Finally, Peri Acid and its isomer Laurent's Acid are separated from the batch filtrate under specific pH conditions.

2. Reactions:

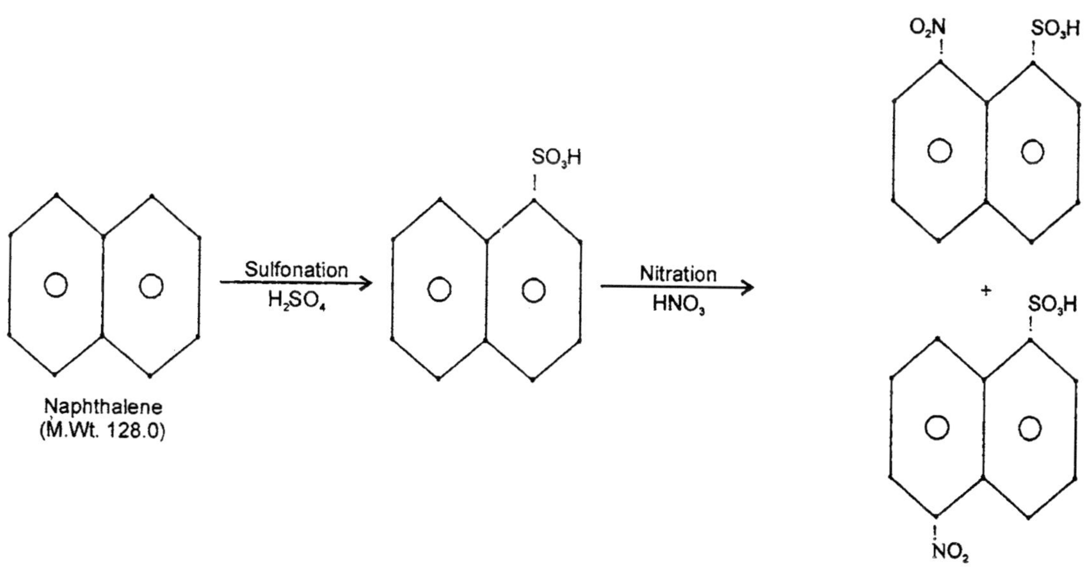

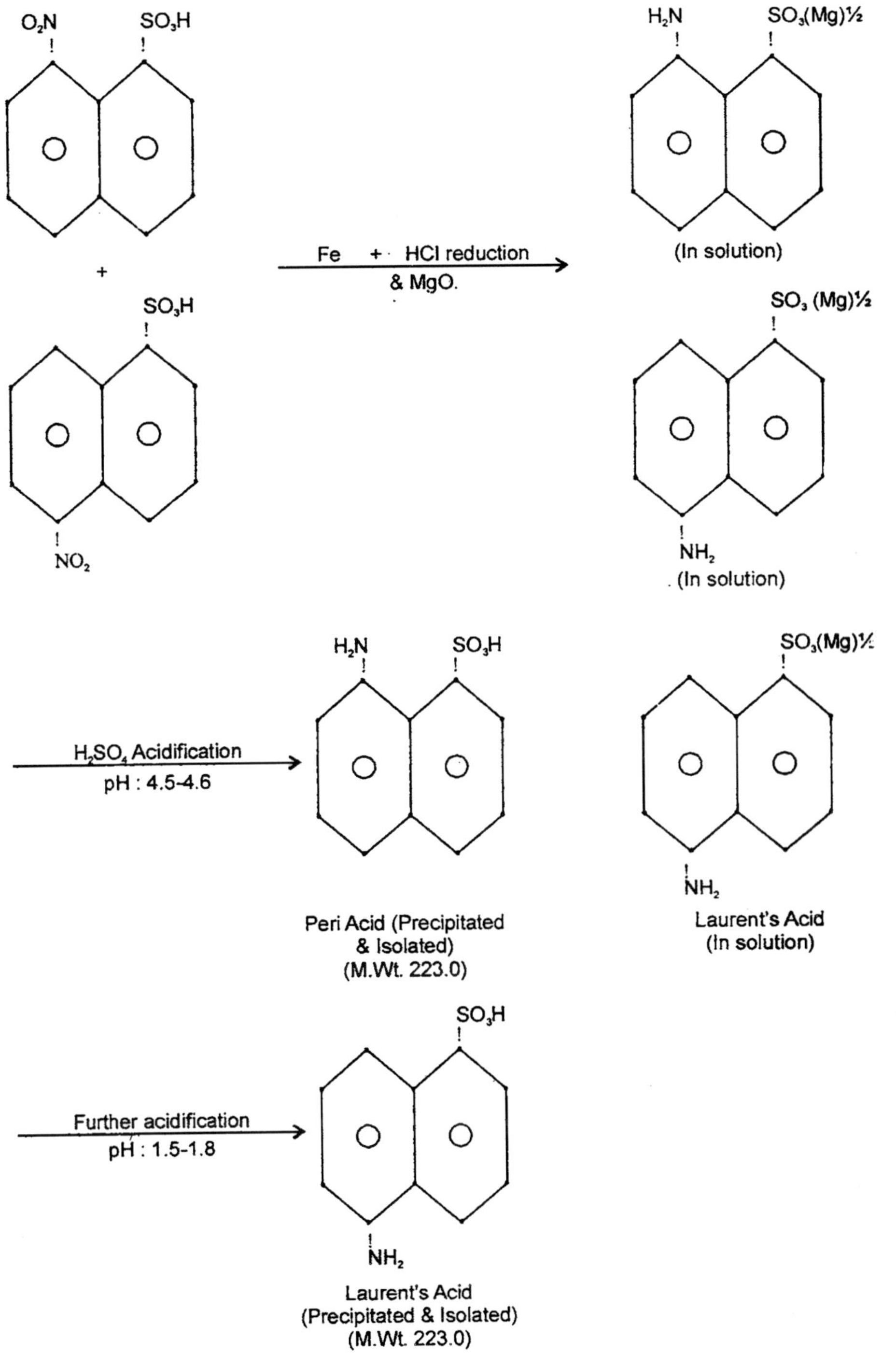
O2N
SO3H
+
SO3H
NO2
Fe + HCl reduction
& MgO.
H2N
SO3(Mg)½
(In solution)
SO3 (Mg)½
NH2
(In solution)
H2SO4 Acidification
pH : 4.5-4.6
H2N
SO3H
SO3(Mg)½
NH2
Peri Acid (Precipitated
& Isolated)
(M.Wt. 223.0)
Laurent's Acid
(In solution)
Further acidification
pH : 1.5-1.8
SO3H
NH2
Laurent's Acid
(Precipitated & Isolated)
(M.Wt. 223.0)

3. Process-Procedure:

3.1. Sulfonation

Tool : A 1500 L capacity C. I. jacketted Kettle equipped with an Anchor-type C. I. Agitator driven by a 5 H. P., 3-Phase, closed electric motor connected with a Reduction-gear giving two speeds of 60-40 r. p. m. to the Agitator. The Kettle is further provided with a Pressure-gauge, a Vent line connected through an exhaust, a Temperature-Recorder-cum-Control (TRC) and a jacketted Blow-out line leading to the Drowning and Neutralisation Kettle.

Check tool.

In the clean, dry, Reaction Kettle,

Charge

730.0 Kgs of as is 98% Sulfuric Acid corresponding to

715.4 Kgs of real Sulfuric Acid.

Start agitator at 40 r. p. m. speed and slowly add in through the manhole

30.0 Kgs of broken Ice lumps.

Stir for 5 minuts.

Draw a sample of the acid and determine its strength; it should be 94 ± 0.2% w/w Sulfuric Acid.

Note the temperature of the acid.

Circulate chilled water in the kettle jacket and cool the acid to 10-15°C.

Drain down water from the jacket.

Then under high speed agitation at 60 r. p. m., charge through manhole

320.0 Kgs of flakes or well ground real Naphthalene powder (M. Wt. 128.0) during about 2 hours.

Close manhole of the Kettle and confirm that the Vent line is open.

Note temperature of the charge; it may rise upto 20°C. Stir the charge for 1 hour at 20-22°C.

Then slowly raise the temperature to 53-55°C over a period of 1 hour by carefully injecting a controlled passage of steam in the Kettle jacket.

Stir the Sulfonation for 3 hours, strictly at 53-55°C.

The sulfonation is now complete.

Note : Near the completion of sulfonation, the mass tends to become very thick. In this case, take care to see that the agitator does not suffer any damage due to excessive load.

At the end of sulfonation, to reduce heavy load on the agitator, add to the batch

60.0 Kgs of as is 98% Sulfuric Acid to thin down the mass. Finally, cool the sulfonated batch to 30°C, first by circulating ordinary water and then chilled water in the kettle jacket.

The volume of the sulfonated batch will be about

600 L at 30°C.

Keep the stirring on at 60 r. p. m. speed.

3.2. Nitration

To the sulfonated batch held at

30°C in the above step,

charge under stirring at 60 r. p. m. speed

25.0 Kgs of as is 62%w/w Nitric Acid corresponding to

15.5 Kgs of real Nitric Acid in a thin stream.

This initial addition of Nitric Acid solution further thins down the thick viscous mass. Cool the charge slowly during about 1 hour to 22-23°C by circulating chilled water in the Kettle jacket.

Then charge slowly, in a thin stream,

230.0 Kgs of as is 62% w/w Nitric Acid corresponding to

142.0 Kgs of real Nitric Acid over a period of 7 hours maintaining the nitration temperature strictly at 22-23°C.

During nitration, the nitration mass thins out at first, but tends to thicken towards the end of reaction. After the addition of Nitric Acid is over, stir the batch further for 3 hours strictly at 22-23°C.

The nitration is now positively complete.

Volume of the batch will be about

775-800 L at 22-23°C.

3.3 Drowning and Neutralisation

Tool An 8,000 L capacity M. S., Rubber-lined and Brick-lined Kettle equipped with a gate-type, M. S. R. L. Agitator driven by a 5 H. P., 3-Phase, closed electric motor connected with a Reduction-gear to give 40 r. p. m. speed to the Agitator.

The Kettle is further provided with a direct steam line, a Pressure-gauge, a temperature-recorder, an exhaust line for nitrous gases leading to the Scrubber, a manhole and a 3-inch bottom-outlet line leading to the gypsum Nutsche Filters.

Check tool.

In the clean Kettle,

Charge

1800 L of treated water or Gypsum-wash water from the previous batch.

Start agitator and charge

90.0 Kgs of real Magnesium Oxide as powder (M. Wt. 40.0). Receive into this, the nitrated batch from the above step during about half an hour.

Note temperature of the drowned batch.

Heat with direct steam to 75°C.

Stop steam.

Then neutralize the acidic batch charge by slowly adding to it a slurry of

650.0 Kgs of real Calcium Carbonate powder (M. Wt. 100.0) prepared separately in

650 L of treated water, during about 2-3 hours.

Take care of foaming during neutralization. Note pH of the neutralized batch; it should be in the range of 7.0-7.3.

If the pH is still lower, add a corrective quantity of extra Calcium Carbonate slurry, as required, to get the pH in the said range.

Stir the neutralized batch for 3 to 4 hours without temperature-control, maintaining the pH in the said range.

Measure temperature and record.

Then steam the neutralized batch again to 75°C. Measure volume of the batch and record:

It will be about

3800 L at 75°C.

The batch is now ready for filtration.

Filtration of Neutralised Batch:

Tool : Two gypsum Nutsche filters.

Each Nutsche is R. C. C., Rubber-lined and Brick-lined equipped with an up-and-down movable rack-agitator and a gypsum side-line to the alkaline Sewer.

Pump the neutralized batch in the two nutsches at 75°C. Collect the batch filtrates from both the nutsches in the Acidification Kettle described under the following step.

Measure volume of the batch filtrate and record.

Now wash the gypsum cakes in both the nutsches with hot water of 75 °C in lots and collect the clear washings along with the main batch volume comes to 4,000 L.

During washing of the gypsum cakes, churn the cakes using the Nutsche agitator.

Finally, wash the gypsum cakes in each nutsche, in the similar way, with

1,000 L of hot water of 75 °C in lots and reserve 1,800 L of these final washings for use in the Drowning and Neutralisation operation of the succeeding batch. Analyse a sample of the filtered batch liquor for total Nitro compound content.

Total conversion accounted collectively in the batch filtrate and final gypsum washings is around 90-95% of theory on the initial charge of real naphthalene.

Press and suck the Gypsum cakes well.

Draw a representative sample of the gypsum cakes and analyse it or accounting the loss of Nitro compound in the cakes.

Weight of Gypsum wet cakes,

= Approx, 2200-2500 Kgs containing about

950-1100 Kgs of dry solids.

Wash down the Gypsum wet cakes straight to the alkaline sewer from the side-lines of Nutsches at the end of washing.

3.4. Acidificatin of Nitro Filtrate

Tool : A 6,000 L capacity M. S., Rubber-lined, Tile-lined Kettle equipped with a gate-type M. S. R. L. agitator driven by a 5 H. P., 3-Phase, closed electric motor connected with a Reduction-gear to give 40 r. p. m. speed to the agitator. The Kettle is further equipped with a direct steam line, a Vent line, a Pressure-gauge, a Temperature-recorder, a manhole and a blow-out line leading to the Reduction Kettle.

The main batch filtrate and the initial gypsum washings from the above step are received in this Kettle.

Check volume of the total batch liquor; it will be

4,000 L at around 65 °C.

Start agitator and add to it

10.0 Kgs of 80%w/w Acetic Acid solution.

Continue stirring until the temperature of the batch liquor falls to 50 °C.

Maintain the batch liquor between 45-50 °C until it is charged up in the Reduction Kettle.

3.5. Reduction

Tool : A 10,000 L capacity M. S., Rubber-lined and Brick-lined Kettle equipped with a Fingure-type Agitator driven by a 7.5 H. P., 3-Phase, closed electric motor connected with a Reduction-gear to give 60/40 speeds to the agitator.

The kettle is further equipped with a direct steam line, a Vent line connected through exhaust to a descent Condensor, a Pressure-gauge, a Temperature Recorder-cum-Control, a manhole and a 3-inch bottom-discharge line leading to the Filter Press.

Check tool.

In the clean Reduction Kettle,

Charge

900 L of wash-water of Iron-Sludge from the previous batch.

Start agitator at 40 r. p. m. speed and charge

730.0 Kgs of Iron borings.

Note : Quality of Iron borings is very important.

Best results are obtained with Cast Iron borings of 175-200 mesh. Coarse iron gives slow reduction.

Then charge

20.0 kgs of as is 30% w/w HCl corresponding to

6.0 kgs of real HCl.

Heat the charge with direct steam to 93-95°C during one hour.

Stir at 93-95°C for 15-20 minutes more to allow the iron to be satisfactorily etched.

Now change the speed of agitation to 60 r. p. m. Then charge in it slowly the acidified nitro compound solution from step 3.4. during about 2 to 2½ hours at such a rate that the reduction temperature is maintained in the range of 94-98°C throughout, as a result of the heat of reduction.

Maintain the reduction temperature in the said range by injecting direct steam, as necessary. During reduction, test for soluble iron as follows:

Test : Spot a drop of reduction sample on a piece of Whatman No. 1 filter paper and cross-spot the clear run-out with a drop of 2.5% Sodium Sulfide solution. A black line developed at the junction of the two spots indicates sufficient soluble iron.

A light-dark or grey line denotes insufficient soluble iron.

If soluble iron in the bath is less as indicated by the above test, correct it by adding a further quantity of iron borings or 30% HCl, as may be necessary, till the test is satisfied.

Confirm sufficiency of soluble iron by above test and continue the reduction further.

Generally, the reduction of first bath in a series requires about 50 Kgs of Iron borings and 12 kgs of 30% HCl as extra charges.

The reduction foams badly.

After the charging of Nitro filtrate is over, continue the reduction for 3 hours at 95-98°C.

The reduction is now complete.

The reduction temperature must be maintained in the said range, throughout the course of reduction.

Test a batch sample for completion of reduction as follows:

Test : Draw 200 ml of the batch sample in a 600 ml beaker. Stir and add sufficient magnesium Oxide powder to obtain a distinct red test on B. Y. paper (pH about 8.0). Filter the precipitated iron through a Buchner funnel and suck. Do not wash the filter cake with water.

(a) Measure 25 ml of the above filtrate in a 500 ml beaker. Add to it 50 ml of water and 10 ml of 30% HCl. Cool to 10°C with some ice and titrate under stirring with a Standard N/5 Sodium Nitrite solution to a 3-minute and point of KI-Starch indicator papers.

Note the burette reading and call it "X".

(b) Measure 25 ml of the same sample filtrate in a 500 ml beaker, add to it 50 ml of water, 5 gms of Zinc dust and 15 ml of 30% HCl. Heat to boil for 10 minutes under slow stirring.

Stop heating, filter through a small Buchner funnel and wash the filter cake with about 50 ml of water. Add to this filtrate 10 ml of 30% HCl, some ice to cool to 10°C and titrate with a Standard N/5 Sodium Nitrite solution to a 3-minute end point on KI-Starch indicator papers. Note the burette reading and call it "Y".

Now, if X is equal to Y, the batch reduction is supposed to be complete. And If Y is greater than X, the batch reduction is regarded as partially incomplete.

If the reduction of batch is still partially incomplete as revealed by the above test, add to the batch corrective quantity of Iron borings and 30% HCl (About 25 Kgs of Iron borings and 5 kgs of 30% HCl) and test again as above after one hour reaction at 94-98°C. Generally, extra quantity of Iron is not required. Confirm the completion of reduction.

Then charge in the batch, as required, about

55.0 Kgs of real Magnesium Oxide as powder (M. Wt. 40.0) so as to obtain the batch pH at 8.0.

Stir for half an hour at 93-95°C and confirm for a distinct red test on B. Y. papers and the batch pH at 8.0.

Confirm the absence of soluble iron in the batch with the following test.

Test : Place a drop of the batch sample on a piece of Whatman No. 1 filter paper. Cross-spot the clear run-out with a drop of 2.5% Sodium Sulfide solution. No black or brown line should be produced at the junction of the two spots. This denotes complete precipitation of soluble iron.

Then add to the batch

10.0 Kgs of activated Carbon and stir for 45 minutes at 90-95°C.

Measure volume of the batch; it will be about

6,000 L at 90-95°C.

The batch is now ready for filtration.

Filtration of precipitated Irion:

Filter the batch hot at 90-95°C through a 1000 L capacity Filter Press having plate and frame dimensions of 48″ × 48″.

Pump the clear batch filtrate in the Separation Kettle described under the following step.

Wash the Press initially with hot water of 80°C in lots and pump these initial washings into the Separation Kettle holding the main batch filtrate to make up the total volume of batch filtrate to 7,000 L.

Finally, wash the Iron sludge in the press further with

1,000 L of hot water of 80°C and preserve this final washing for the use in the reduction of succeeding batch.

Dump the iron sludge from the Press.

Weight of wet Iron sludge

= Approx. 2500-2700 Kgs.

= Approx. 1000-1100 Kgs total solids.

Wash down the wet iron sludge to alkaline drains.

Seperation of 1:8 and 1:5-Isomers

3.6. Precipitation of Peri Acid: (1:8-Isomer)

Tool A 10,000 L capacity M. S., Rubber-lined Kettle equipped with a Spade-type M. S. R. L. Agitator driven by a 7.5 H. P., 3-Phase, closed electric motor connected with a Reduction-gear to give 40 r. p. m. speed to the agitator. The Kettle is further provided with a direct steam line, a vent line connected through an exhaust, a Pressure-gauge, a Temperature-recorder, a manhole and a bottom-discharge line leading to the Nutsche filter.

		The batch filtrate together with initial Press washings from the above stop containing
		1:8 and 1:5-Naphthylamine Sulfonic acids are collected in this Kettle.
		Check volume of the solution; it will be
7,000	L	at around 80°C.
		Start agitator and stir to cool to 75°C.
		If the temperature is lower than 75°C,
		Steam the batch to 75°C.
		Now stop heating.
		Then slowly add to it, as required, in a thin spray
185.0	Kgs	of as is 40%w/w Sulfuric Acid corresponding to
74.0	Kgs	of real Sulfuric Acid to adjust the batch pH strictly in the range of 4.5-4.6.
		Note : Do not add excessive acid to run down the pH below 4.5 at any time. However, if it still happens, raise the batch pH to 6.0 by adding Liquor Ammonia and then readjust the pH back in the said range by adding judicious quantity of Sulfuric acid solution.
		Stir the batch for 1-1/2 hours without temperature-control.
		Note pH and confirm it to be in the said range;
		If not, adjust.
		The batch temperature falls to about 65-68°C.
		The 1:8-Isomer i. e. Peri Acid precipitates out under thee conditions.
		Measure batch volume ; it will be about
7,200	L	at 65-68°C.
		The batch is now ready for filtration.
		Filter the precipitated Peri Acid at 65-68°C through an M. S. R. L. Product Nutsche having 7.4 Sq. Meters basket area.
		Wash the Nutsche cakes with about
800	L	of ice-cold water, till the washing become almost colourless.
		Preserve the M. L. and washings in the Kettle described under step 3.7. for the isolation of 1:5-Isomer, i. e. Laurent's Acid.

Press the Peri Acid wet cake in the Nutsche and suck well.

Dump the Nutsche.

Contrifuge the Peri Acid wet cakes collected from the Nutsche.

Dump the Centrifuge, weigh the wet cakes and record.

Note : If desired, the precipitated Peri Acid may be directly filtered through a Centrifuge and washed as specified. The additional Nutsche filtration may be avoided, this way.

Collect the residual M. L. received from the centrifuge filtration. Add this M. L. to the main batch M. L. retained for Laurent's Acid isolation as stated above. Dump the centrifuged cakes, weigh and record. Draw a representative sample of the centrifuged Peri Acid wet cakes from above and analyse it for real Peri Acid content by a Nitrite titre.

Calculate and record yield of Peri Acid.

3.7. Precipitation of Laurent's Acid: (1:5 Isomer)

Tool : A 10,000 L capacity M. S., Rubber-lined Kettle equipped with a Spade-type M. S. R. L. Agitator driven by a 7.5 H . P., 3-Phase, closed electric motor connected with a Reduction-gear to give 40 r. p. m. speed to the agitator. The Kettle is further provided with a direct steam line, a vent line connected through an exhaust, a Pressure-gauge, a Temperature-recorder, a manhole, an immersion Cooling coil of lead and a bottom-discharge line leading to the Nutsche Filter.

The M. L. and Washings containing Laurent's Acid retained after the filtration of Peri Acid are preserved in this Kettle as stated earlier.

Measure volume of this solution; it will be about

7,300 L at 45-50°C.

Adjust temperature of this solution to 60°C by steaming.

Stop heating.

Check pH of the solution and record; it will be in the range of 4.4-4.6.

While stirring, charge in it slowly in a broad spray, as required, about

110.0 Kgs of as is 40%w/w Sulfuric Acid solution corresponding to

44.0 Kgs of real Sulfuric Acid so as to adjust the pH in the range of 1.5-1.7.

Stir for half an hour without temperature-control and confirm the pH in the said range.

Cool the batch to 45°C by circulating chilled water through the Cooling coil.

Stir the batch for 1-1/2 hours at 43-45°C.

Measure pH and confirm to be in the said range.

Laurent's Acid precipitates out under these conditions. Measure Volume of the batch; it will be about

7,400 L at 43-45°C.

The batch is now ready for filtration.

Filter the precipitated Laurent's Acid through an M. S. R.L. Product Nutsche having 7.4 S. meters basket area.

Suck the nutsche cakes and wash with

200 L of ice-cold water.

Release the M. L. and washing to acid-drains.

Press the nutsche cakes and suck well.

Dump the nutsche, collect the wet cakes and centrifuge.

Discard the M. L. to drains derived from centrifuging.

Note : If desired, the precipitated Laurent's Acid may be directly filtered through a Centrifuge and washed as stated. The additional Nutsche filtration may be avoided, this way.

Dump the centrifuged cakes, weigh and record.

Draw a representative sample of the centrifuged cakes

From above and analyse it for real Laurent's Acid content by a Nitrite titre.

Calculate and record yield of Laurent's Acid.

4. Yield:

4.1 1-Naphthylamine-8-Sulfonic Acid (Peri Acid)

Weight of centrifuged cakes,

= Approx. 400-450 Kgs.

= Abour 245.0 Kgs of real Peri Acid (M. Wt. 223.0).

= 44% of Theory on the charge of real Naphthalene.

Dry the centrifuged wet cakes of Peri Acid in vacuum at 70°C.

4.2. 1-Naphthylamine-5-Sulfonic Acid (Laurent's Acid):

Weight of centrifuged cakes,

= Approx, 160-180 Kgs.

= about 88.0 Kgs of real Laurent's Acid (M. Wt. 230.0)

= 15.8% of Theory on the charge of real Naphthalene.

Dry the centrifuged wet cakes of Laurent's Acid in vacuum at 70°C.

5. Raw Materials Usage Data:

	Name of Raw Materials	*M. Wt.*	*Quantity Real (Kgs)*	*Moles*	*One Mole Ratio*
i)	Naphthalene, powder	128.0	320.0	2.5000	1.0000
ii)	Sulfuric Acid	98.0	715.0	7.3000	2.9200
iii)	Sulfuric Acid	98.0	58.8	0.6000	0.2400
iv)	Nitric Acid	63.0	158.1	2.5095	1.0038
v)	Magnesium Oxide	40.0	90.0	2.2500	0.9000
vi)	Calcium Carbonate	100.0	650.0	6.5000	2.6000
vii)	Acetic Acid	60.0	8.0	0.1333	0.0533
viii)	Iron Borings (C. I.)	55.8	730.0	13.0824	5.2329
ix)	HCl	36.5	6.0	0.1644	0.0657
x)	Magnesium Oxide	40.0	55.0	1.3750	0.5500
xi)	Sulfuric Acid	98.0	74.0	0.7551	0.3020
xii)	Sulfuric Acid	98.0	44.0	0.4489	0.1795
xiii)	Activated Carbon	--	10.0	–	–

24

Manufacture of Schaeffer's Acid (2-Naphthol-6-Sulfonic Acid)

1. Principle:

Beta Naphthol is sulfonated with concentrated Sulfurric Acid in about 83% average concentration in a medium of Glauber Salt solution. The sulfonated Product resulted is practically all Schaeffer's Acid containing only traces of other Isomers. It is then isolated and centrifuged.

2. Reactions:

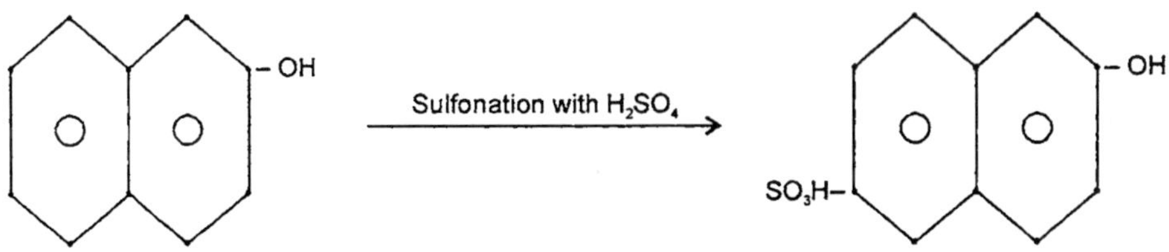

Beta Naphthol (M.Wt. 144.0) — Schaeffer's Acid (M.Wt. 224.0)

3. Process-Procedure:

3.1. Sulfonation

Tool A 2000 L capacity C. I. Jacketted Kettle with an M. S. jacket and equipped with an Anchor-type Agitator driven by a 5 H. P. Electric motor connected with a Reduction-gear to give 40 r. p. m. speed to the agitator. The Kettle is further provided with a Temperature-recorder, a Pressure-gauge and a bottom-outlet line leading to the Isolation Kettle.

		Check tool.
		In the clean Sulfonation Kettle
		Charge
660.0	Kgs	of as is 98% Sulfuric Acid corresponding to
646.8	Kgs	of real Sulfuric Acid.
		Start agitator.
		Start circulation of chilled water in the Kettle jacket and add in slowly
180.0	L	of treated water, allowing the final temperature to keep at 30°C.
		Now under continued cooling through jacket slowly charge in the batch.
66.0	Kgs	of Anhydrous Glauber Salt (M. Wt. 142.0) and cool the batch charge to 20°C.
		Then, holding the batch temperature between 20-25°C charge
430.0	Kgs	of real Beta Napthol (M. Wt. 144.0) as flakes during about 3/4 to 1 hour.
		A thick cream-coloured slurry will be obtained.
		Stir or half an hour at 20-25°C and charge in it
380.0	Kgs	of as is 98% Sulfuric Acid corresponding to
372.4	Kgs	of real Sulfuric Acid.
		Then charge
42.0	Kgs	of Anhydrous Glauber Salt.
		Drain down the chilling water from the Kettle jacket. Inject a passage of steam in the jacket and heat the batch as fast as possible to 90°C.
		At 90°C, the batch slurry thins down and almost a complete solution is obtained. However, after about first one or two hours' stirring, precipitates appear to be falling out. Stir the batch strictly at 90-92°C for ten hours. The sulfonation is complete by now.
		At the end of sulfonation, add an extra quantity of
55.0	Kgs	of as is 98% Sulfuric Acid corresponding to
54.0	Kgs	of real Sulfuric Acid to thin down the resulting thick slurry of the sulfonated mass.
		The batch is now ready for Isolation.
		Measure volume of the sulfonated batch; it will be about
1150-1200	L	at 90°C.

3.2. Drowning and Isolation

Tool : A 5,000 L capacity Lead-lined M. S. Kettle with an M. S. Jacket equipped with a 7.5 H. P. electric motor and an Anchor-type, lead-lined M. S. Agitator having variable speed.

The Kettle is further equipped with a Temperature-recorder, a Pressure-gauge and a bottom-discharge line leading to the Product Nutsche.

Check tool.

In the clean Isolation Kettle,

Charge

2,800 L of treated water.

Start stirrer at 40 r. p. m. speed and receive into this the sulfonated batch from the above step during about 30-45 minutes. The temperature of the drowned bath mass comes to about 55-60°C.

At 60°C, slowly add to the batch

250.0 Kgs of Anhydrous Glauber salt during about one hour.

Heat the batch to 90°C during one hour.

The drowned mass keeps in a slurry from until the temperature reaches to 85°C.

Ultimately, the mass starts thinning down and goes into an almost complete solution when the temperature reaches to 90°C.

Now reduce the stirring speed to 20 r. p. m. Stop jacket-steam and stir the batch to cool.

The product starts crystallising out at about 80°C. Then charge in the batch

20.0 Kgs of as is 30%w/w Hydrochloric Acid (Colourless, Iron-free) corresponding to

6.0 Kgs of real Hydrochloric Acid (M. Wt. 36.5).

Stir for three hours at 75-80°C.

The batch slurry thickens with increased crystallization.

Stir to cool the batch to 65°C without circulating water in the jacket.

At 65°C, start circulation of cooling water in the Kettle-jacket and cool to 55°C during about two hours. Finally, circulate chilled water in the jacket and further cool the batch to 30°C during about two hours. Measure batch volume and record; it will be about

4,000-4,200 L at 30°C.

The batch is now ready for filtration.

Filtration:

Before starting filtration, prepare the Cake-washing solution by dissolving

15.0 Kgs of Anhydrous Glauber Salt,

5.0 Kgs of Ammonium Sulfate and

0.25 Kgs of E. D. T. A Disodium Salt in

300.0 L of treated water.

Filter the batch at 30°C through a wooden, brick-lined Nutsche or a Wooden Nutsche.

Suck the cakes as much as possible.

Analyse a sample of the M. H. for %age loss of Schaeffer's Acid.

Pass the M. L. to the Treatment System through Acid-drain.

Break the vacuum in the Nutsche.

Then pour over the well-sucked

Nutsche-cakes, a First lot of

150.0 L from the Cake-washing solution prepared as above. Allow the washing solution to slowly percolate and leach into the product cakes. After a while. Apply vacuum and suck down the washing as much as possible from the Product cakes.

Again break the Nutsche vacuum.

And then, give the second washing with the final lot of

150.0 L of the washing-solution, exactly as above.

Suck the washed Product-cakes very well.

Analyse a sample of the combined Cake-washings for %age loss of Schaeffer's Acid.

Dump the Nutsche.

Then quickly centrifuge the Nutdche cakes without any prolonged standing.

Weight the centrifuged Wet cakes, analyse a sample for Schaeffer's Acid content and record Yield.

4. Yield:

Weight of the centrifuged Wet Cakes,

= Approx 750-775 Kgs

= 500.0 Kgs of real Schaeffer's Acid (M. Wt. 224.0)

= 74.7% of Theory on the charge of Beta Naphthol.

Notes: (i) With a regular batch campaign of well controlled operations, the yield of the Product can be raised o as high as 80% of Theory.

(ii) Iron content in a sample of the centrifuged wet cakes is determined by the conventional Method of Colour-comparision in Nessler's tubes against a Standard, using Potassium thiocyanate as the developer of coloured Iron complex.

5. Raw Materials Usage Data:

	Name of Raw Materials	M. Wt.	Quantity Real (Kgs)	Moles	One Mole Ratio
i)	Beta Naphthol	144.0	430.00	2.9861	1.0000
ii)	Sulfuric Acid, Real	98.0	646.80	6.6000	2.2102
iii)	Glauber Salt, Anh.	142.0	66.00	0.4648	0.1556
iv)	Sulfuric Acid, Real	98.0	372.40	3.8000	1.2726
v)	Glauber Salt, Anh	142.0	42.00	0.2958	0.0991
vi)	Sulfuric Acid, Real	98.0	54.00	0.5510	0.1845
vii)	Glauber Salt, Anh.	142.0	250.00	1.7606	0.5896
viii)	Hydrochloric Acid, Real	36.5	6.00	0.1644	0.0550
ix)	Glauber Salt, Anh.	142.0	15.00	–	–
ix)	Ammonium Sulfate	132.0	5.00	–	–
xi)	E. D. T. A. Disod. Salt	–	0.25	–	–

25

Manufacture of Rhoduline Acid (Di-J-Acid)

1. Principle:

Sodium Salt of J-Acid is treated with a Conc. Solution of sodium Bisulfite and reacted at 106-109°C on the principle of Buchrer Reaction till the Rhoduline Acid formation is complete. The crude Product is then salted out and purified.

2. Reactions:

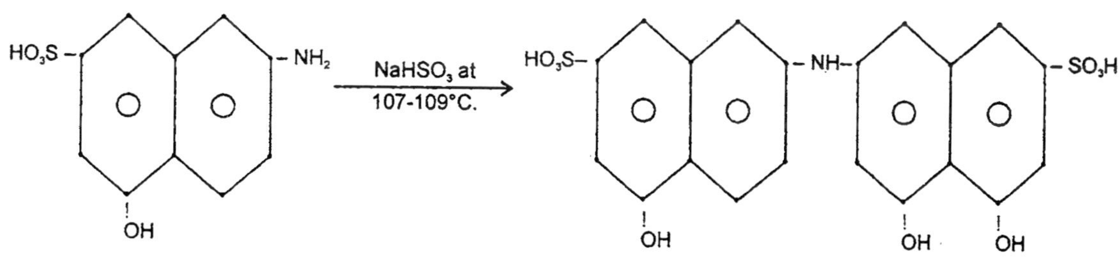

J-Acid, (M. Wt. 239.0) Rhoduline Acid, (M.Wt. 461.0)

3. Process-Procedure:

In a 5000 L.M.S. Lead-line jacktted vessel equipped with a spade-type M. S. Lead-lined agitator having variable speed and a reflux condenser, charge

475.0	L	of water. Stir and add
585.0	Kgs	of as is 95% Sodium Bisulfite equivalent to
555.0	Kgs	of Real Sodium Bisulfite (M. Wt. 104).
		Stir to slution under a speed of 35 r. p. m.
		Adjust the volume to
1000.0	L	by adding requisite amount of water.
		Add

164.0 Kgs of as is 50%w/wt Caustic Soda solution equivalent to

82.0 Kgs of Real Caustic Soda and follow with an addition of

16.0 Kgs of Real Sodium Sulfite (M. Wt. 126).

Then charge slowly, during about 1-1/2 hours, controlling the foaming,

478.0 Kgs of Real J-Acid (M. Wt. 239) as Sodium Salt.

J-Acid having a titre value of Min. 65% as real

J-Acid (M. Wt. 239) can be employed.

Stir half an hour after addition of J-Acid is over.

Note pH; it should be between 4.3-4.5.

Measure volume; it will be about

1400 L Adjust the volume to

1700.0 L with water.

Close the vessel.

Keep the roof-vent valve over the Condensor partially open Start cooling water on the Reflux Condensor. Then heat the Batch to 107-109°C during about 4 hours using 2 Atoms. Steam in the jacket. The reaction foams strongly at the beginning. The temp of the out-going Gas through the reflux-condensor should be adjusted below 85°C by controlling the flow of water on the Condensor-jacket. This temp is recorded by a Thermometer fixed in the vent-pipe Once the temp of 107-109°C is attained, reduce the stirrer speed to 25 r. p. m. and stir the reaction for 35 Hours at 107-109°C.

Stop steam on the jacket and stir the batch to cool to 90°C. Record volume.

Dilute the Batch with simultaneous passage of water and live steam to

3400.0 L at 90°C.

Note pH; it should be preferably 6.6-or within the range of 6.3-6.6.

Then reduce the stirring speed to 20 r. p. m. and add uniformly during about 1-1/2 hours

760.0 Kgs of Vacuum Salt at 90°C.

Stop heating and stir to cool to 30°C during about 5 hours circulating raw water in the jacket, if necessary.

Filter the Bath on Filter Press.

Discard the Mother Liquor to drains.

Air the Press to no drip.

Dump the Press, weigh the Wet Cakes and record.

The wet Product is thixotropic and jelly-like in nature.

Weight of Wet Cakes = Approx. 1300-1500 Kgs.

4. Purification

In a 12,000 L. M. S. Jacketted Rubber-lined vessel equipped with a spade-type agitator having two speeds, an exhaust-line to heat-exchanger and leading to SO_2-Scrubber, charge

900.0 L of water and steam to 50°C. Stir at 35 r. p. m. speed and add the wet cakes crude Rhoduline acid obtained in the previous step. Stir half an hour 50-55°C.

Then by adding about

75.0 Kgs of 30%w/wt HCl, as required, adjust the pH between 2.8.3.1.

Cover the manhole with a wooden lid and steam to 65 degC with direct steam. Evolution of SO_2 gas takes place which goes out the SO_2 scrubber.

Stir at 65°C until no more SO_2 gas is evolved.

This takes about 1-1/2 hours.

Finally, dilute the batch to about

8000.0 L volume at 70°C by alternate passage of live steam and water. The Product should just dissolve giving a Milky solution. It is preferable to keep the volume low to 7500 L if the dissolution can be done with lesser water.

Then add, as required, about

5-6.0 Kgs of Real Caustic Soda as a 50% w/wt solution to adjust the pH between 5.9-6.1.

Measure the volume and record. Reduce the stirring speed to 20 r. p. m.

Add to the system Common Salt a rate of

25.0 Kgs per every 100 L volume during about 1-1/2 hours at 65°C and stir for 1 hour more at this temp. Approximately

2000.0 Kgs of Common Salt may be required.

Then stop heating.

Circulate raw water in the jacket and start cooling. When the rate by jacket cooling slows down, insert in the vessel a portable cooling Coil and cool the Batch to 20°C over a period of 8 hours.

Draw a sample of precipitated Product, filter in Laboratory and check the Baume of filtrate.

The filtrate should show 23°C Baume. If it is less, adjust the Baume of the filtrate to the said limit by adding requisite amount of Salt. Confirm the completion of precipitation of the Product.

Filter the Batch at 20°C on Product-Nutsche or a Filter Press.

Air the Press and suck to no drip.

Dump the Press.

Centrifuge the Wet Cakes at 800 r. p. m.

The wet Centrifuged product is slightly buff to light-brown in colour.

5. Yield:

Weight of Centrifuged Wet Cakes = Approx. 725-750 kgs

= Approx. 345 Kgs of Real Rhoduline Acid (M. Wt. 461).

= 75 of Theory

6. Raw Materials Usage Data:

	Name of Raw Materials	*M. Wt.*	*Quantity Real (Kgs)*	*Moles*	*One Mole Ratio*
i)	Sodium Bisulfite	104.0	555.00	5.3370	2.6685
ii)	Caustic Soda	40.0	82.00	2.0500	1.0250
iii)	Sodium Sulfite	126.0	16.00	0.1270	0.0635
iv)	J-Acid	239.0	478.00	2.0000	1.0000
v)	Vacuum Salt	–	760.00	–	–
vi)	HCl, 30%w/wt.	36.5	22.50	0.6165	0.3082
vii)	Caustic Soda	40.0	6.00	0.1500	0.0750
viii)	Common Salt	–	2000.00	–	–

26

Manufacture of G-Salt (2-Naphthol-6:8-Disulfonic Acid)

1. Principle:

Beta-naphthol is sulfonated with a mixture of Sulfuric Acid and Oleum. The sulfonated mass is drowned in water and the G-salt precipitated as Potassium salt by adding Potassium chloride to the G-Acid solution. Finally, the isolated G-salt is filtered out and the filter cakes centrifuged. R-Salt is isolated from the M. L. left after G-salt filtration.

2. Reactions:

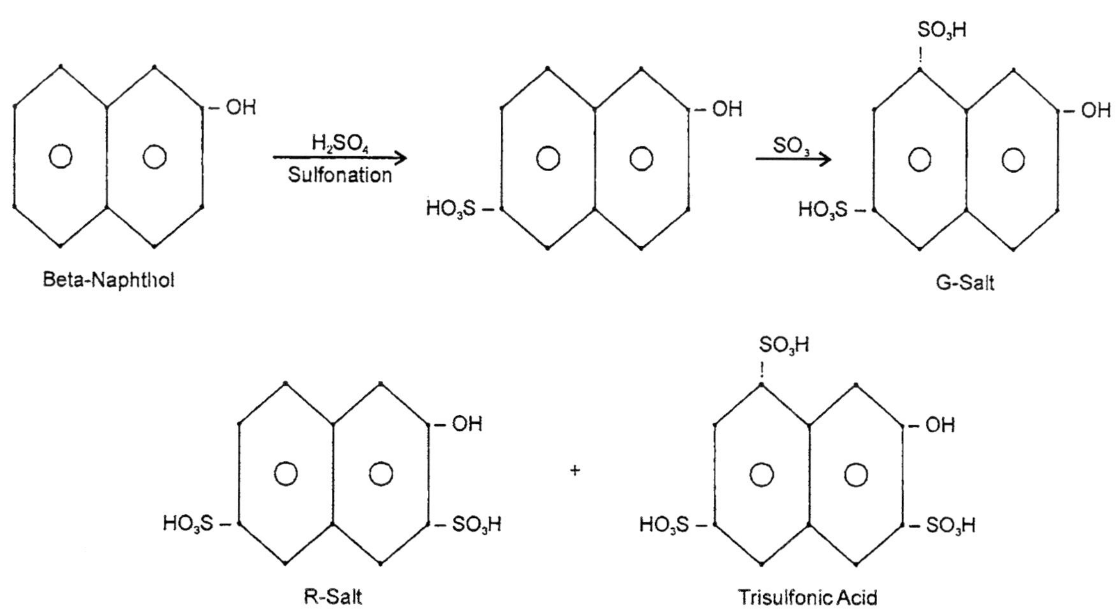

3. Process-Procedure

3.1. Sulfonation

Tool A 1000 L capacity jacketed C. I. sulfonator equipped with an Anchor-type Agitator having two speeds and a temperature-recorder.

Check tool.

Charge in the Sulfonator

140 Kgs of as is 98% Sulfuric Acid corresponding to

402 Kgs of Real Sulfuric Acid. Circulate cooling water in the jacket and open exhaust line start agitator and stir at a speed of 20 r. p. m.

When the acid cools to 20-22°C,

Start uniform charging of

364 Kgs of Real Beta-Naphthol (M. Wt. 144) so as to complete the charging over a period of about 3-1/2 to 4 hours. The temperature, during charging should not be allowed to exceed 45°C. Preferably, the temperature should be maintained in the range of 40-45°C.

Check volume; it will be about

535 L at 45°C.

Continue jacket cooling to cool the charge to 32°C during about 2 hours.

Throttle the vent valve to allow a minimum escape of fumes.

At this stage, almost all the Beta-Naphthol goes into solution.

Now increase the speed of agitator to 40 r. p. m. Weigh in a suitable Measuring Vessel

650 Kgs of as is 25% Oleum corresponding to

162.5 Kgs of real Sulfur Trioxide (M. wt. 80).

Make the addition of Oleum to the above charge in the following pattern.

140 Kgs Oleum @ 35 Kgs/hour	=	4 Hrs.
360 Kgs Oleum @ 100 Kgs/hour	=	3 Hrs.
150 Kgs Oleum @ 75 Kgs/hour	=	2 Hrs.
650 Kgs Oleum. Total period	=	9 Hrs.

During the addition of Oleum, control the rise in temperature of the sulfonation mass in such a way that the temperature may rise to 52° maximum in the final hour of addition. During sulfonation, the reaction mass tends to become every thick.

Check this condition during reaction and if the load on stirrer increases due to thickening, thin the mass by adding to the charge a maximum quantity of

10 Kgs of as is 98% sulfuric Acid,

without stopping the stirrer.

Then heat the batch carefully in the following manner.

Maintain temp at 53-5°C for	2 hours
Heat from 55 to 65°C in	3 hours
Heat from 65 to 75°C in	3 hours
Stir at 75-78°C for	6 hours
Total Period:	14 hours

The order of heating should be carefully followed. Measure volume: it will be about

825 L at 78°C.

Test for completion of sulfonation as follows:

Test : Draw a sample of the sulfonated mass.

Drop about 10 gms of the sample in a 500 ml beaker containing about 200 ml of water while stirring with a rod.

Add to it gradually solid Soda-bi-car powder until the solution imparts a faint red test on B.Y. papers. Watch for any undissolved residue in the solution.

If the solution is clear, the sulfonation is complete.

Usually, the sulfonation is complete by now.

At the end of sulfnation, the mass becomes very thick and so it is advised to drown without any delay.

3.2. Drowning

Tool At 4000 L capacity M. S. L. L. jacketed Vessel equipped with a Gate-type Agitator having variable speed and a temperature recorder.

Check tool.

Half-an-hour before completion of Sulfonation, charge in the Drowning Vessel G-salt wash water from the previous batch and by adding requisite quantity of treated water make up volume to

2100 L. Start agitator. Set the agitator at 40 r. p. m. speed. Heat the water to 45-50°C by jacket steam.

On completion of sulfonation, receive the sulfonated batch into the Drowning Vessel over a period of about 1-1½ hours. On completion of drowning, the temperature of the drowning mass rises to 90-95°C.

If less, heat externally by jacket steam to 90-95°C.

Then add to the batch

535 Kgs of real Potassium Chloride (M. Wt. 74.5).

slowly during about 2 hours.

Slower charging of KCl gives bigger crystals of the product and a quicker filtration. Measure volume of the batch an record.

Adjust the volume of the batch of

3250 L at 90°C by adding to it requisite quantity of hot water of 90°C. Correct form of G-salt precipitates out at this stage.

Now heat the batch further to 100-102°C by jacket steam (or by heating coils) during about 1 hour. Stir the batch at this temperature for ½ hour.

Then stop heating and cool the batch slowly in the following pattern.

100°C to 70°C during	1 hr a 40 r. p. m. speed,
70°C to 50°C during	2 hrs at 20 r. p. m. speed, and
50°C to 20°C during	3 hrs at 20 r. p. m. speed.
Total cooling period :	6 hours.

Finally, stir the batch at 20 ± 1°C

for 6 hours and proceed with filtration.

3.3. Filtration

Filter the batch at 20°C through

a product Nutsche, suck the cake well, collect M. L. in a 4000 L capacity M. S. R. L. Nutsche-egg.

Volume of M. L. will be about

2600 L. Wash the wet cake in Nutsche twice, each time with

50 L of cold water of 15°C.

Collect these washing separately and preserve for reuse in the Drowning step of the succeeding batch.

Suck the cakes well.

Dump the Nutsche and centrifuge the cakes, each load for 50 minutes at 800 r. p. m.

Weigh the wet cakes. Analyse a sample and record yield.

Note : The total M. L. collected at the end of separation of G-salt is retained for the recovery of R-salt as described in the Appendix.

4. Yield:

Weight of the Centrifuged wet cakes,

= Approx. 750-775 Kgs having about 65-67% Real G-salt content (M. Wt. 304).

= 500 Kgs of Real G-salt (M. Wt. 304).

= 65% of Theory on Beta-Napthol charge.

5. Raw Materials Usage Data:

	Name of Raw Material.	*M. Wt.*	*Quantity Real (Kgs)*	*Moles*	*Mole Ratio*
1.	Beta-Naphthol	144.0	364.0	2.5270	1.0000
2.	Sulfuric Acid, 98%	98.0	402.0	4.1020	1.6230
3.	Oleum, 25%	—	650.0	—	—
	SO_3	80.0	162.5	2.0310	0.8035
4.	Potassium chloride	74.5	535.0	7.1810	2.8410

6. Recovery of R-Salt:

Refer Appendix for recovery of R-Salt from M. L.

27

Manufacture of 4-Nitro-4′-Amino-Stilbene-2:2′-Disulfonic Acid

1. Principle:

4:4′-Dinitro-Stilbene-2:2′-disulfonic Acid is reduced in aqueous medium under controlled conditions of temperature and pH using Sodium Hydrosulfide free from Sodium Sulfide. The reduction liquor is acidified with Sulfuric Acid, SO_2 gas is boiled off, the liquor basified again, screened out and the product finally precipitated from the clear filtrate by acidification. The bath is filtered and the wet product evaluated for real content. The product is used or sold as such, in wet form.

2. Reactions:

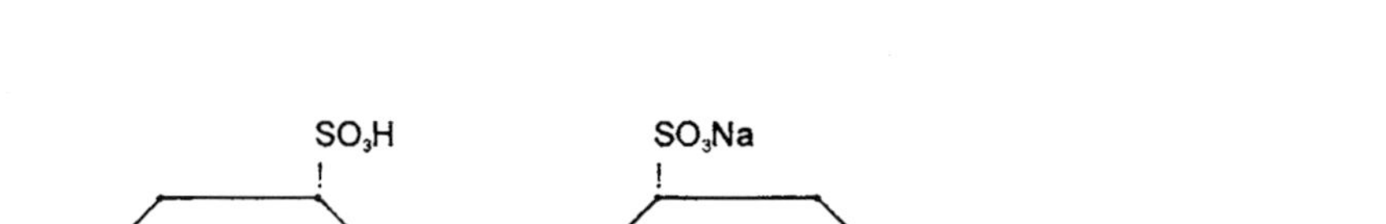

DNSDSA
(M.Wt. 430.0)

DNSDSA
(M.Wt. 400.0)

3. Process-Procedure

3.1. Calculation of Sodium Hydrosulfide Charge

Calculate total charge of NaSH necessary for the batch reduction in the basis of 1.34 Moles of NaSH for 1.0 Mole of 4:4′-Dinitro-Silbene-2:2′-disulfonic Acid. The NaSH liquor (About 30-35% NaSH content) to be used for the reduction should be freshly analysed for:

(i) % NaSH content and

(ii) % Na_2S content.

The percentage Na_2S content of the liquor is calculated in terms of % NaSH by multiplying the % Na_2S by 0.718.

Thus, the total % NaSH content of the liquor is represented by the sum of % of NaSH actually determined in the liquor and the % NaSH factor deduced by calculating % Na_2S as % NaSH.

To convert Na_2S to NaSH, add 1.10 of real Sodium Bicarbonate for every 1.0 kgs of Na_2S present in the NaSH liquor.

From above, determine the total % NaSH content in the NaSH liquor and calculate the actual quantity of NaSH liquor to be taken for use.

Also calculate the quantity of Sodium Bicarbonate necessary for conversion of Na_2S to NaSH as explained above.

3.2. Preparation of Sodium Hydrosulfide Charge

Tool A 2,000 capacity Wooden vat equipped with a paddle-type wooden Agitator driven by a 3 H. P., 3-phase, closed Electric motor connected with a Reduction-gear to give 40 r. p. speed to the agitator. The Vat is further equipped with a top wooden cover having a charging hopper on the cover, a Thermometer, four wooden baffles fixed on the inside wall of the Vat and bottom-discharge line leading to the Reduction Kettle.

Check tool.

In the clean Vat,

Charge

750.0 L of treated water (at 30°C).

Start agitator.

Then charge in it

313.0 Kgs of as is 34% Nash liquor corresponding to a quantity of

106.5 Kgs of total real NaSH (M. Wt. 56.0).

Note : The NaSH liquor used analyses as, 33.0% real NaSH (M Wt. 56.0) and 1.4 % real Na_2S content (M. Wt. 78.0).

Now, since 1.4% Na_2S corresponds to

(1.4×0.718) = 1.00% NaSH, the total percentage NaSH content of the liquor is (33.0 + 1.0) = 34.0 %.

At this rate, the above batch charge of 313.0 Kgs of as is 34% NaSH liquor contains 103.3 Kgs of actual NaSH and 4.38 Kgs of actual Na_2S. 4.38 Kgs of Na_2S on conversion produces 3.14 Kgs of NaSH. Conversion of 4.38 Kgs of Na_2S to NaSH requires (4.38×1.1) = 4.82 Kgs of real Sodium Bicarbonate.

followed with,

5.0 Kgs of real Sodium Bicarbonate (M. Wt. 84.0) to convert the Na_2S in the charge to NaSH.

Stir to solution.

The solution is now free from Sodium Sulfide content.

Then charge an extra quantity of

20.0 Kgs of real Sodium Bicarbonate (M. Wt. 84.0).

Stir to solution.

Measure volume of the solution and adjust it to

1,100.0 L at 30°C with treated water.

Stir the solution for 30 minutes and clarify through a Clarifier Pres or Nutsche Filter.

The clear solution is now ready for use in Reduction.

3.2. Reduction

Tool A 6,000 L capacity C. I. jacketted Kettle (Instead of C. I. Vessel, a Wooden Vat or an M. S. Kettle can be used for Reduction) equipped with an anchor-type, C. I. Agitator driven by a 5 H. P., 3-phase, closed electric motor connected with a Reduction-gear to give 40 r. p. m. speed to the agitator. The Kettle is further equipped with a Temperature-Recorder, a Pressure gauge, a Vent line connected with an exhaust fan to drive off flue gases above the plant roof, a manhole and a blow-out line leading to the SO_2 boil-off Kettle.

Check tool.

In the clean Reduction Kettle,

Charge

2,000.0 L of treated water.

Start agitator and charge

610.0 Kgs of real 4:4′-Dintro-Stilbene-2:2′-Disulfonic Acid (M. Wt. 430.0) as disodium Salt, followed with

70.0 Kgs of Borax, $Na_2B_4O_7.10H_2$ (M. Wt. 382.0).

Heat the charge to 70°C with jacket steam.

Measure volume of the charge and adjust it to

3,000.0 L at 70°C with more of treated water, as required. Stir for one hour at 70°C with jacket steam. Analyse a sample of the charge for % DNSDSA content (M. Wt. 430.0). Also record pH.

Calculate total real DNSDSA in the charge.

Add uniformly during two hours, half the volume of the NaSH solution from above through a fine Sprinkler, holding the temperature strictly in the range of 68-72 degC. Note pH; it must be in the range of 9.2-9.7.

Then add uniformly through the Sprinkler, half of the remaining NaSH charge during two hours holding the reduction temp at 68-72°C and the pH strictly in the range of 9.2-9.7.

Finally, add the last quarter, that is the balance NaSH charge through the Sprinkler during three hours maintaining the reduction temperature at 68-72°C an the pH strictly in the range of 9.2-9.7.

During the progress of reduction, if the pH tends to rise above 9.7. correct it by adding Borax. Usually, the pH does not shift. But at times if it so happens, Borax has to be added to hold the pH within the specified range.

In such case, the requirement of Borax hardly exceeds 5-7 Kgs for the entire batch.

After all the Hydrosulfide solution is added up, stir the reduction at 68-72°C one hour maintaining the pH between 9.2-9.7.

During reduction, occasionally test for the absence of Sulfide by cross-spotting a drop of tha reduction mixture with a drop of neutral Ferric Chloride solution on a piece of filter paper. Development of no black line or colour at the junction of the spots Genotes the absence of Sulfide.

Draw a reduction sample at the end of 1 hour stirring and analyse for:

(i) % Degree of reduction,

(ii) % Dinitro-Stilbene-Disulfonic Acid content (unreduced) and

(iii) % Diamino-Stilbene-Disulfonic Acid content.

N. B.: For analysis, refer Appendix I.

The expected result of analysis of the firs sample is:

(i) About 90% Degree of reduction,

(ii) 8-10 unreduced Dinitro-Stilbene-Disulfonic Acid.

(iii) 0.3 to 0.4% Diamino-Stilbene-Disulfonic Acid.

As revealed by the analysis of the sample, if the degree of reduction is 90%, add to the batch corrective quantity of NaSH solution to raise the reduction degree to 96%, which is considered to be the ideal limit.

Addition of corrective NaSH Solution, if required:

For each additional degree of reduction, i. e. for every 1% of residual reduction, prepare a solution by dissolving

2.0 Kgs of real Sodium Hydrosulfide (M.Wt. 56.0) and

0.4 Kgs of Sodium Bicarbonate in

25.0 L of water.

Add this solution to the reduction mass slowly during one hour at 68-72°C and holding the pH between 9.2-9.7.

Stir the reduction for 1/2 hour at 68-72°C.

Then draw a second sample of reduction mass and analyse.

The analysis of second sample will usually show:

(i) 96% Degree of reduction,

(ii) 4-5% Dinitro-Stilbene-Disulfonic Acid (unreduced) and

(iii) 0.3-0.6 % Diamino Stilbene Disulfonic Acid.

When analysis of the batch sample is as above, the reduction is regarded as ideal and satisfactory.

When the reduction is complete, measure batch volume; it will be about

4,400.0 L at 68-72°C.

3.3. Decomposition of Sulfides and SO_2 boil-off

Tool A 7,000 L capacity M. S. Lead-lined, jacketted Kettle equipped with an anchor-type M. S. L. L. Agitator driven by a 5 H. P., 3-phase, closed Electric motor connected with a Reduction-gear to give 40 r. p. m. speed to the agitator. The Kettle is further provided with a Temperature-Recorder, a Pressure-gauge, a Vent line connected to the SO_2-Scrubbing system, a direct steam line, a manhole and a bottom-discharge line leading to the Screening Press.

Check tool.

In the clean Kettle, transfer the reduced batch from the Reduction Kettle.

Wash the Reduction Kettle with about 200 L of treated water and receive the washings along with the batch mass. Measure volume; it will be about

4,600.0 L at about 60-65°C.

While stirring, add to it cautiously, as required, about

232-235 Kgs of as is 98% Sulfuric Acid to adjust the pH in the range of 2.5-3.0. the Sulfuric Acid addition takes about ½ hour. SO_2 gas starts coming out as a result of H_2SO_4 addition which is vented to the Scrubber.

Heat the batch to 98-100°C with direct steam

(35 lbs pressure) and start the boil-off of SO_2 gas. Maintain a positive blue test on Congo Red papers, throughout the SO_2 boil-off.

Add more Sulfuric Acid, if necessary, to maintain the Congo test and pH in the said range. SO_2 boil-off takes about 3-4 hours at 98-100°C. After the SO_2 boil-off is over, cool the batch to 85-90°C by circulating ordinary water in the Kettle jacket.

Note : 30% HCl can as well be used in place of 98% H_2SO_4, subject to selection of Kettle of appropriate material of construction.

3.4. Basification and Screening

Add to the batch slowly and cautiously, controlling the foaming, Soda Ash powder to adjust the pH of the batch between 7.8-8.3. Take care of brisk foaming while adding Soda Ash. To case the foaming problem, partly Caustic Soda solution, 40%w/w, may be added for initial neutralization of the batch and the residual neutralization may be accomplished by the addition of Soda Ash.

The solution, at the end of neutralization, becomes almost clear with some sulfur particles floating in suspension.

The basification requires about

150-160 Kgs of Soda Ash powder, provided all Soda Ash is used. Check and confirm the pH to be in the range of 7.8-8.3. Stir for 15-20 minutes and again confirm the pH. Measure volume; it will be about

5,200.0 L at 85°C (without foam).

Then clarify the solution through a Clarifier Press or Nutsche.

Pump the clear batch filtrate to the Isolation Vessel described in the following step.

Wash the Clarifier with about 200 liters of hot water of 90°C and collect the washings along with the batch filtrate in the Isolation Vessel.

3.5. Isolation and Filtration

Tool A 7,000 L capacity, M. S. Lead-lined, jacketted Kettle equipped with an anchor-type, M. S. L.L. Agitator driven by a 5 H. P., 3-Phase, closed Electric motor connected with a Reduction-gear to give 30 r. p. m. speed to the agitator. The Kettle is further equipped with a Temperature-Recorder, a vertical Vent line, a Pressure-gauge, a manhole and a bottom-discharge line leading to the Filtration tool.

Check tool.

In the clean Isolation Kettle is pumped the clear batch filtrate along with washings from the previous step. Start agitator.

Measure temperature and pH and record.

Heat the batch to 85°C with jacket steam and adjust the volume to

5,500.0 L as 85°C.

While stirring, add as rapidly as possible, about

110-115 Kgs of as is 98% Sulfuric Acid to adjust the pH in the range of 1.5-2.3. A strong-blue test on Congo Red papers should be obtained.

Stir for 15-20 minutes and check pH.

Then circulate water in the jacket and cool the batch to 30-32°C.

The product precipitates out as sparingly soluble Acid-Sodium salt.

Measure volume: it will be about

5,700.0 L at 30-32°C.

Stir the batch for 1/2 hour at 30-32°C.

The isolation is now complete and the batch ready for filtration.

Filter the product through Filter Press or Nutsche filters. Suck the M. L. and give about 150 L vessel-wash of water on the cakes.

Suck well and centrifuge the wet cakes.

Discard the M. L. and washing to acid drains.

Note : Filtration of the isolated batch through Centrifuge is very convenient and have recommended.

4. Yield:

Weight of centrifuged batch cakes,

= Approx. 850-900 kgs containing about 56-59 % real NASDSA (M. Wt. 400.0)

= 500 Kgs of real 4-Nitro-4′-Amino-Stilbene-2:2′-disulfonic

Acid (M. Wt. 400.0)

= 88.0% of Theory on the real charge of 4:4′-Dinitro-Stilbene-2:2′-Disulfonic Acid

The product obtained is as Mono-Sodium Salt.

It contains less than 1% DASDSA and about 5-6% of unreduced DNSADA.

5. Raw Materials Usage Data:

	Names of Raw Materials	*M. Wt.*	*Quantity Real (Kgs)*	*Moles*	*Mole Ratio*
i)	4:4′-Dinitro-Stilbene-2:2′-Disulfonic Acid (As Sod. Salt)	430.0	610.0	1.4186	1.000
ii)	Sodium Hydrosulfide	56.0	106.5	1.9018	1.3406
iii)	Sodium Bicarbonate	84.0	25.0	–	–
iv)	Borax	382.0	70.0	–	–
v)	Sulfuric Acid	98.0	230.3	2.300	1.6565
vi)	Soda Ash	106.0	160.0	1.5094	1.0640
viii)	Sulfuric Acid	98.0	112.7	1.1500	0.8106

Appendix I

Process Analysis:

Analyse the reduction mass samples for determining the degree of reduction as desci ibed below:

Procedure:

Pipette 50 ml of the reduction sample into a 500 ml Volumetric flask. Dilute to the mark with distilled water and mix well. Call this solution as "V".

1. Pipette 100 ml of solution "V" in an 800 ml beaker. Add 200 ml of water and 25 ml of conc. HCl. Boil off SO_2, which may take about half an hour, maintaining the volume. Filter the SO_2-free solution through filter paper in a clean 1-L beaker and wash the paper well with more of water.

   To the clear filtrate, add enough ice to cool to 15°C. Add 5 gms of sodium Bromide and titrate with a Standard N/10 Sodium Bromide and titrate with a Standard N/10 Sodium Nitrite solution to a 3-minute distinct end-point on KI-Starch papers.

   Note the titre reading and call it "X".

2. Pipette 50 ml of solution "V" in a 600 ml beaker. Add to it 25 ml of conc. HCl and 100 ml water. Heat to gentle boil under stirring and add Zinc dust in small portions until the solution does not decolourise further. This takes about 4-5 gms Zinc dust.

   Maintain the volume of solution in the beaker during reduction. Boil further for 15-20 minutes. Then add to it 25% w/vol Caustic soda solution to get a distinct red test on B. Y. papers.

Filter the solution hot through buchner funnel in a clean flask. Wash the zinc cake with 100 ml of hot water of 90°C. Remove the zinc cake from the funnel and reslurry it with 100 ml of hot water Filter this cake-wash through buchner and mix the clear filtrate of washing with the main filtrate.

Transfer these filtrates into a clean 1 L beaker an ad enough ice to cool to 15°C.

Add to it 25 ml of Conc. HCl and 5 gms of Sodium Bromide. Titrate the mixture, under stirring, with a standard N/10 Sodium nitrite solution to a 3-minute distinct end-point on KI-Starch papers.

Note the titre reading and call it "Y".

Then,

Degree of Reduction = (X/Y × 100) %.

28

Manufacture of N-Phenyl H-Acid

(Total time cycle 100-120 hours)

Tool : MSGL Autoclave, 2000 L, 5 H. P. Motor, 100 RPM, anchor type stirrer.

		Check tool.
		Charge
400.0	L	water. Start stirrer.
		Charge
266.0	Kgs	Real H-Acid in 1 hour by breaking lumbs.
		Then charge
400.0	Kgs	Aniline Oil and
110.0	Kgs	40% w/w H_2SO_4.
		Measure Volume: 1350 L
		If not adjust by water
		Close manhole.
		Heat to 130°C slowly in 3 hours.
		The pressure may reach between 44 to 60 lbs/in^2.
		Stir at 130-132°C for 38 hours.
		Watch for any leakage in the autoclave.
		Cool to 95°C.
		Transfer the mass to steam Distillation Kettle.
		During the transfer remove a sample from the sample point (an arrangement of sample point in the blow line) for w/w/ H-Acid unreacted in the condensed mass.
		Give 100 L hot water wash (80°C) by vacuum, and blow the wash in the distillation Kettle.

Then open the Vessel and check for any left material.

Cool the vessel at Room Temperature.

Distillation:

Tool : M. S. Vessel 3000 L
7.5 H. P. Motor.
100 1bs Steam.

Check tool.

Charge

63.0 Kgs Real NaoH as 45% w/w solution into the already received back-wash of 500 Lts. Then receive the reaction mass along with washes. During the period test on B. Y. paper should be positive; if not add some more NaoH lye to get the test.

pH: 8.0-9.0

Adjust volume to

2150-2200 L with water.

Heat to 90°C by jacket steam, then to 100°C by live steam. Continue steam distillation till the distillate is clear.

Collect the distillate.

Approx: 2000 L

Seperator:

Distillate = 2000 L

Charge

20% w/v Salt to the distillate.

Stir for 1/2 hour.

Allow to settle and separate water and Aniline layer (upper one) Aniline collected will be 220 Kgs, and should be free from turbidity and oily drops.

Check the volume, it should be

2200 L if not adjust it by water. Transfer to clarification Vessel. Take 500 L hot water wash and blow to the clarification vessel.

Clarification Vessel MSRLTL:

Check the vessel.

Receive the mass from the Distillation Vessel with wash. The temperature is 80°C.

Adjust pH to 1.5-1.7 by adding 40% w/w/ H_2SO_4 About

100-120 Kgs will be required.

		Add
7.0	Kgs	Darco and
7.0	Kgs	Hyflosupercel
		Stir for ½ hour. Clarify through filter to isolation vessel.
		Wash with about 400 L water of 75-80°C and filter.
		Again take 400 L water of 75-80°C, and pass through filter, take in the Distillation Vessel as back-wash. Total volume of clarified solution will be 2800-2900 L, if not adjust.
		Isolation:
		Tool : 4000 L MSRL, with 5 H. P. motor 15 RPM
		Receive the clarified solution from clarified vessel (2800-2900 L)
		Check sample for:—
		1. Clarity
		2. % Real N-Phenyl H-Acid.
		Heat to 75°C.
		Adjust pH to 1.0 ± 0.1 by
		adding 40% w/w H_2SO_4
		About
100-110	kgs	of 40% w/w H_2SO_4 will be required. Check pH
		Add at 70-75°C
		22% w/v VCF salt in 2 hours, uniformly with
		intermittent stirring (very important)
		Cool slowly to 35°C in 6-8 hours.
		Check °Be of Ml, it will be 20°Be
		Filte in Nutsche (MSLR)
		Receive the batch and suck well.
		Wash the product with brine solution
		(400 L water + 80 Kgs solt and acidity with H_2SO_4 to pH 1.0)
		Suck dry
		Dump an centrifuge.
		Wet cakes: about 560 kgs.

Yield:

246-250	kgs	real N-Phenyl H-Acid
		73% theory on charge of H-Acid.

Raw Material Usage Data:

Name of Raw Material	*M. Wt.*	*As is Qty. Kgs*	*Real Qty. Kgs.*	*Moles*	*Mole Ratio*
H-Acid	319.0	–	266.00	0.8338	1.0000
Aniline Oil	93.0	–	400.00	4.3010	5.1580
H_2SO_4' 40%	98.0	110.00	44.00	0.4489	0.5384
NaOH, 44%	40.0	63.00	27.22	0.6930	0.8311
H_2SO_4 40%	98.00	110.00	44.00	0.4489	0.5384
Salt	–	200.00	–	–	–
VCF Salt	–	638.00	–	–	–
Darco	–	7.00	–	–	–
Hyflosupercel	–	7.00	–	–	
Aniline Recovered	93.00	–	220.00	–	–

29

Manufacture of 2:6 Dichloro Para Nitro Aniline

GIL Vessel

		Charge OR Suck in
2700		litres Ethylene dichloride
		= 3400 Kgs.
400		litres water.
		Volume will be 3100 litres.
		Slowly in about one hour
		Charge
180	Kgs	Real Para Nitro aniline as centrifuged W. C. of above 95% N.V.
		Charge
8	Kgs	Sulfur Powder.
		Stir about 30 to 45 mts.
		Heat to 50°C in 1/2 hour by jacket steam.
		Then slowly in
		5 hours add
200	Kgs	chlorine Gas maintaining temp. by cooling water in jacket.

Stir for 2 hours at 50°C and check the sample for completion of Chlorination. Unreacted P. N. A. should not be more than 0.5 to 1.5 kgs. If more, then add corrective Cl_2 and recheck.

If unreacted P. N. A. is within the limited, then

Stir the batch for 4 hours at 50°C. Slowly then cool to Room Temperature by cooling water in jacket in 4 hours.

Then circulate chilled water in jacket and cool to 10°C in 3 hours. Stir the batch at 10°C for 2 hours. Then dilute the batch by adding 500 Kgs ice + 500 litres water. Temperature should be 10°C Volume will be about 4400 to 4500 litres. Filer the batch on PN.

Suck m. l. in a batch egg

Where previously 300 to 315 Kgs.

Real NaOH is taken. Then give 150 litres EDCL wash to the W. C. Then give 4 × 200 litres water washes to the W. C.

Suck well.

Then give 20 Kgs Soda Ash dissolved into 100 litres water wash to the W. C.-suck well.

Give 2 × 200 litres water washes.

W. C. should be acid free.

Suck dry well. Dump W. C.

And centrifuge the W. C.

Dry the W. C. at 80°C in a

Vac Dryer in 24 to 30 hours.

Dump dry material.

Wt. of dry material = 190 to 205 Kgs.

Yield-191 Kgs from

180 Kgs Real P. N. A.

Product will be of

Dull or light yellow coloured in crystalline form.

Purity—98.5 to 99.5%

Recovery of EDCL

Heat to 90°C with jacket steam (30 p. s. i.) and then pass live steam.

Steam distill EDCL.

This will take about 9 hrs.

Reuse EDCL for the next batch.

Recovery: 92 to 95%

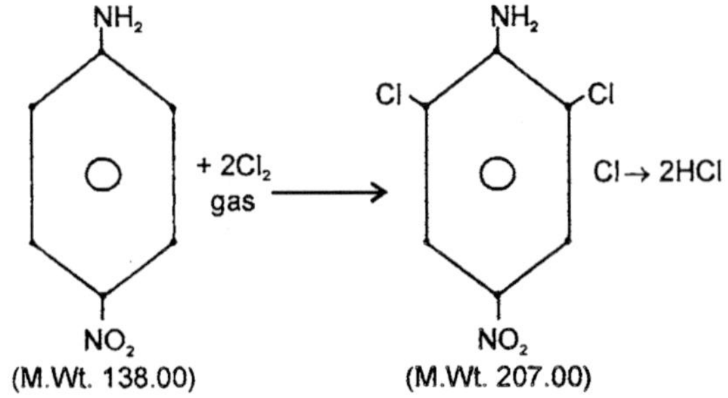

30

Manufacture of 2:6-Dibromo-4-Nitroaniline (2, 6 dibromo 4 PNA)

Raw Material:

Para Nitro Aniline	–	75 Kg
Sodium bromide NaBr	–	100 Kg
Chlorine	–	90 Kg
T. R. O. (Turkey Red Oil)	–	7 Liter.

Process:

Close the bottom valve of reactor. Charge 1000 liter H_2O charge 75 Kg of PNA and 100 Kg of NaBr add 7 liter T. R. O. steam is passed on, continue the agitation, then pass the Cl_2 at bottom at the rate of 14 kg/hour. About 90 Kg of Cl_2 is passed. Then take out sample from reactor and check M. P. If M. P. is O. K. stop passing Cl_2.

If it is not O. K. then pass some more Cl_2 and send sample for checking M. P. If it O K. drawn in a tank and keep it for one hour. Filter in F. P. make it acid free.

Charge again in drowning kettle containing 500 liter H_2O and 10 Kg Soda Ash strand filter it.

Wash it till neutral.

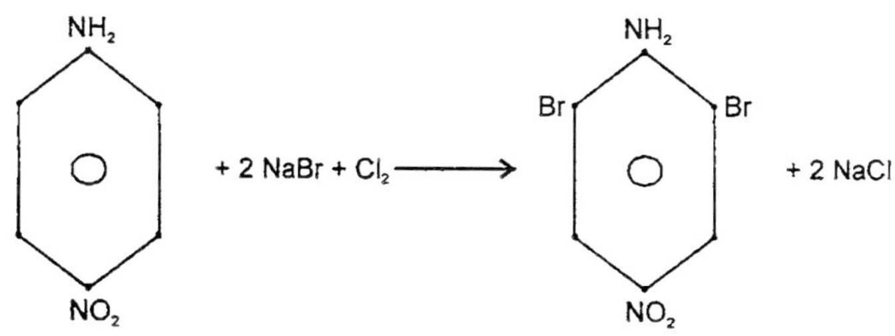

31

Manufacture of (Chlorotoluic Acid-CLT-Acid) 6-Chloro-m-Toluidine-4-Sulfonic Acid

		MV : 101
680	Kg	Weigh 98% w/w Sulfuric Acid in the weigh-tank and receive it in the M. V.
		Volume: 370 Ltrs.
		Add this Sulfuric Acid into V 1281.
		Sufonation & Chlorination:
		Check tool.
		Check all valves.
		Check all Accssories.
		Suck
500	Kgs	Toluene.
		Start stirrer.
		Open valve of condenser.
		Start cooling water in condenser. Hat toluene to 95°C with 25 lbs
		Steam in jacket in one hour.
		Stop heating
		Charge
680	Kgs	98% H_2SO_4 slowly in 1 hour from MV 101 maintaining temperature at 95-100°C
		(If required, use CW or steam in jacket)
		After complete addition of Sulfuric Acid, raise temperature to 116-118°C in about 1 hour.

Stir at 116-118°C for 8 hours.

Cool to 90°C in about 1 hour.

Note volume: 925 Ltrs.

Send a sample to Lab for completion of Sulfonation Report.

If the sulfonation is reported incomplete, add corrective amount of 98% H_2SO_4 in Lots of 54 Kgs. Heat to 116-118°C and stir for 3 hours at 116-118°C. Recheck the sample.

If the sulfonation is complete, cool the batch to 50°C. Take

1989 Kgs of 98% w/w sulfuric Acid in weigh tank and charge it into V 1281 through MV, at temperature of 50°C in about ½ hrs. Add

3.2 Kgs Ferric Chloride (anhydrous).

Note Volume : @ 2000 Ltrs.

Note Temp : @ 50-52°C.

Now start passing

396 Kgs Chlorine gas at the rate of 8-9 Kgs/hour through the Chlorine sparge at 50-52°C Chlorine gas addition will take about 15 hours.

(chlorination rate is very important factor)

After Chlorine addition is over, cool to 25°C by CW in jacket.

Pass dry air to remove any unreacted Chlorine and HCl gas, through the bottom for one hour.

Not the Volume: 2100 Ltrs

Blow the batch to V1164 by a direct line.

(V 1282: Time Cycle: 64-68 hours)

V 1164: Nitration

Check tool.

Check all the valves.

Check all accessories.

Receive the batch from V 128 L.

Note the volume and temp.

Volume : @ 2100 Ltrs

Temp : @ 20-25°C.

If not, bring down the temp to 20-25°C with brine in jacket and start the stirrer.

Add ice slowly in 3-4 hours.

Maintain the temp of reaction.

Mass at 25-28°C by external cooling.

Note Volume: about 2200 Ltrs

Stir for 10 minutes.

Start addition of

353 Kgs 98% w/w Nitric Acid (= 234 Ltrs) from S. S portable MV, maintaining temp at 25-28°C.

Note volume after addition Nitric Acid is over

(about 2418 Ltrs).

Total Nitric Acid addition time will be about 7 hours.

Further stir the batch at 25-28°C for 4 hours.

Then blow the batch to V 1333.

(V 1164 Time Cycle: about 21 hours.)

V 1333

Check tool.

Check all valves.

Check all accessories.

Take

3300 Ltrs Water.

Start stirrer.

Charge

1152 Kgs Common Salt and stir to dissolve the salt.

(This gives 22% w/vol brine solution.).

Heat to 95°C by direct steam and adjust the volume to 5250 Ltrs at 95°C.

Add

65 kgs 20% Dedanol OT solution (20% w/v).

Now receive Nitro mass from V 1164 in about 1-2 hours.

Temp rises to 95-98°C.

Note the temp.

Note the volume: about 6900-7200 L

Cool uniformly to 30-32°C in a 4-5 hours with CW and then CHW in head-cool.

Note volume.

Filter the batch on PN 1001/PN 1002

V-1333: Time Cycle-hours.

PN-1001/PN 1002: Filtration

Check the product nutsch.

Receive the batch from V 1333, and suck dry ML in batch egg V 1705/ V 1710.

Wash the wet cake with

4 × 1350 Ltrs of 30% w/volume Brine solution.

Suck dry and dump and centrifuge wet cake.

Yield:

Expected W. C. : 2782 – 3600 Ltrs

% N. V. : 24 – 31.

Real Kgs : 916 Kgs.

= 63% Theory on Toluene.

Batch egg: V 1710

Suck Ml and wash

Volume: 9800 – 10300 Ltrs

% N. V. : 4.0-4.5%

Real : 440 Kgs

V 1328 Reduction:

Check tool.

Check all valves.

Check all accessories.

Charge

3000 Ltrs Water or back wash of previous batch if present. Heat to 92°C by live steam and adjust the volume to about

3500 Ltrs.

Charge

350 Kgs Iron powder and

23 Kgs Acetic Acid.

Stir at 90-95°C for ½ hour

Add

6.7 Kgs Dedanol OT and Stir.

Check pH. It should be 5.5 to 6.0

(Test on MRY paper should be positive and on C. R. paper should be negative).

Charge

1700 Kgs of Nitro wet cakes showelwise in about 3 hours duration keeping temp at 90-95°C.

(1/2 batch Nitro Wet Cake obtained from 500 Kgs Toluene)

When Nitro Wet cake charging is over,

Check the volume and temp.

Volume will be about 4890 Ltrs.

Now raise the temp to 98°C.

Stir at 98-100°C for 5 hours.

(Note: During heating, check the pH of reduction mass. It should be around 5.5 to 6.0 Also the Test of soluble iron must be positive).

Send sample to Lab. For completion of Reduction.

Report: Reduction is complete/incomplete.

If the reduction is reported to be complete.

Add

180-200 Kgs Soda Ash to get a faint test on phenolphthalein paper. Check the pH.

It should be 8.0-8.5.

Add

11 Kgs Activated Carbon and stir for 1 hr at 90-95°C.

(Test of soluble iron should be –ve)

Note volume: About 5500 ± 200 Ltrs.

Blow the batch to FP 1001 initially by short blow leg under stirring.

Insert long blow leg and blow the remaining reduction mass to Filter Press without stirring at 92-95°C.

Take 1350 Ltrs hot water wash of 90°C and blow to FP 1001.

Finally give 2 × 1500 Ltrs hot water of 90°C as back wash.

(V 1328 Time Cycle: 24-28 hours)

FP 1001: Clarification

Check the press.

Receive reduction mass from V 1328 and collect the clarified main liquor into V 1262.

Receive hot water wash of 1350 L and clarify and collect into V 1262 upto volume 6800 Ltrs.

Receive 2 × 1500 Ltrs hot water washed from blower at 90°C, clarify and collect it into V 1262 as back wash.

Air dry the sludge in FP.

Dump and analyse the sludge after weighing it.

Weight of Iron sludge: 1113 Expected

1000-1250 Actual

% N. V. 1.0 to 1.5

Real 10 to 12 Kgs

V. 1262-V1323

Receive the clafiried liquor from FP 1001 in two lots (via short blow leg and blow leg)

Receive hot wash of vessel.

Note volume and temp.

Temp should be 90°C, if not heat to 90°C.

Volume should be 6800 Ltrs.

After receiving 6800 Ltrs of volume analyse the sample for N-Value and real Kgs.

(N. V. 6.0 to 7.5%)

460 to 500 kgs.

Transfer to V 1323 for isolation.

Receive about 3000 Ltrs of hot wash as back wash from FP 1001.

Keep it...for next batch usage in V 1328.

V 1323-Isolation

Check tool

Check all valves

Check all accessories

Receive the main liquor and hot

Water washes from V 1262. Note the volume and Temperature.

Volume should be 6800 ltrs.

If not adjust with water at 60°C.

Start stirrer.

Note temp. It should be 60°C.

Adjust the temp at 60°C either by CW in jacket or with live steam.

Add

9.5 Kgs Sodium Hydrosulfite.

Stir for 1 hour.

		Add
4.0	Kgs	Dedanol OT.
		Now slowly in 1 hour, add
304	Kgs	Hydrochloric Acid 30% w/w, at 60-70°C, to bring pH to 1.0-1.5.
		Stir for 1 hour at 65-70°C
		Check pH again: 1.0-1.5.
		Send sample to Lab for % N. V.
		of M. L. (It should be 0.7 ± 0.1%
		Note the volume: 7000 to 7500 Ltrs.
		Filter at 65°C on PN 1010.
		(V 1323 Time Cycle: 12-14 hours.)
		PN 1010: Filtration:
		Check pH; batchegg, connections required.
		Receive the batch from V 1323.
		Suck ML into suitable batchegg V 17.
		Wash the W. C. with 2 × 750 750 Ltrs hot water
		wash of 50-60°C.
		Suck dry.
		Dump and centrifuge.

Yield:

Net Weight	:	412-442	From 250 Kgs Toluene.
% N. Value	:	65-70	
Real Kgs	:	287	

M. L. in Batchegg V 17

M L x Wash	:	9230 L
% N. V.	:	0.76
Real Kgs	:	70.00

32

Manufacture of 4-4′Diamino Diphenylamine-2-Sulfonic Acid

Step 1: Reduction: V 1328

		In a Reduction Kettle, MSRLT, 10,000£, Finger type
		Charge
2100	L	water and heat to 90°C by live steam and adjust to volume
2900	L	at 90°C. Note temperature and volume.
		Charge
175	Kg	Iron filings or C. I. Power.
		Stir and add
30-60	Kgs	30% w/w Hydrochloric Acid in about 10 minutes. Heat the batch to 100°C. Stir at 100°C for ½ hour. Now charge during about 2.5 hours
200	Kgs	p-nitro Aniline powder (M. Wt. 138.0) maintaining temperature at 90-100°C. Then charge
175	Kgs	another Iron filings of C.I. Powder and
30-60	Kgs	30% w/w Hydrochloric Acid. Stir for 15 to 20 minutes and Charge
200	Kgs	remaining p-Nitro Aniline powder (M. Wt. 138.0) during about 2-1/2 hours.

The test of soluble iron should be checked throughout PNA addition and it should be positive.

If not, add corrective quantity of about 25-30 Kgs of Iron filings or powder and hydrochloric Acid 30% w/w. Check pH (it should be 5.0 ± 0.2) check volume-2900 L to 3200 L. Close manhole and reflux the batch for 4 hours at 100°C. After reflux of 4 hours, analyse the sample for completion of reduction.

If reduction is incomplete, add corrective iron filings and hydrochloric Acid 30% w/w and reflux at 100°C for 2 hours. Recheck for completion of Reduction. If reduction is complete, cool the batch to 90°C.

Charge

100-150 Soda Ash powder in ½ hour.

Stir for 1 hour at 90-95°C. Check and report pH.

(It should be about 9.0 ± 0.2)

Check and note the volume,

2800-3200 L at 90°C.

Blow the batch at 90°C to FP 1001 for removal of Iron sludge.

Wash the Iron sludge with

1000 L × 2 hot water of 90°C.

Time cycle: 18 to 24 hours.

Filter Press: FP 1001 48"-24 frames

Check and reset the Press. Pre-water it after reset. Pre-heat the press with steam well. Receive reduced mass from V-1328 and clarify it.

Receive 1st wash of 1000 L hot water of 90°C and clarify. Complete filtration of main liquor and washes and receive ML + washes in V 1261/v 1262.

Air dry the sludge in Press. Dump the sludge. Weigh and analyse. Weigh of Iron sludge 700 to 900 Kgs (725 Kgs expected)

% w/w N. V. 1.5 to 2.0%

Real Kgs: 10-15 Kgs.

FP 1001: Time Cycle: 12-16 hours.

V-1261/V 1262. SS 8000 L with 12.5 M^2 Condenser

Check the tool.

Check all valves.

Check all accessories.

Receive clarified main liquor of reduction mass from FP -1001.

Note the volume of main liquor. Receive 1st wash from FP-1001 and note the volume and temp.

Final volume of ML + washes will be

4800 L

Adjust volume to 4800 L.

Send sample and analyse the strength of PPD liquor.

PPD liquor: 4800 L to 5000 L.

% w/vol. of PPD: 6.0 to 6.5%

Expected Real: 298.5 Kgs.

290-305 kgs.

Step II: Condensation

V 1261/V 1262

To the above received 4800-5000L of PPD solution (about 298.5 kgs real PPD) of pH 9.0 to 9.2, charge common salt (31.5% w/vol)

Charge

186-240	kgs	Soda Ash powder in hour.
		Temp will be @ 55 to 60°C.
		Note the volume and temp.
		Light brown coloured thin slurry will be obtained. Charge
672.0	Kgs	PCNB-3-Sulfonic Acid as a wet cake.
		Stir.
		Check pH and volume.
		pH should be 9.5 to 10.0 and volume
6400	Ltrs.	

If not, adjust both pH and volume. Start water in condenser and Stir for ½ hour.

Close manhole and heat the batch to 105-107°C in 2-3 hours.

Reflux the batch at 105-107°C for 10 hours.

Cool the batch to 80-85°C with cooling water in jacket.

Open the manhole and send a sample to check completeion of condensation.

(The slurry will be of brownish colour with few insolubles) If the condensation is complete, note the volume and blow the condensed mass to V 1322 at 80-85°C.

V 1261/V1262 Time Cycle: 26-30 hours.

V-1322 10,000£ MSRL

Check tool.

Check all values

Check all accessories.

Receive condensation mass of 80-85°C from V1261/V1262 into V-1322.

Check volume and note it.

6200-6400 L at 80-85°C.

Check and report pH and temp.

Heat to 85°C (if required).

250-270 Kgs Add 98% w/w Sulfuric Acid in about 3 to 4 hours at 80-85°C. to get pH 1.0 ± 0.2.

Stop addition of 98% w/w H_2SO_4

When pH 1.0 + 0.2 is achieved.

A yellowish Khakhi coloured slurry will be obtained. Finally cool the batch to 30-35°C by external cooling. Filter the batch on PN

Check and note volume before filtration.

V 1322: Time Cycle: 18 to 22 hours.

PN 1009/1008 with Carbon tiles

Filtration

Check the nutsche well. Receive the batch from V 132. Suck dry. Wash wet cake with 5 × 1142 Litres water of room temp. Suck dry well.

Dump wet cake.

Expected W. C. : 1190 kgs (1500 to 1800 kgs)

Expected N. V. : 70%

Expected Real : 833 Kgs.

% Theory on PNA : 92.86

% Theory on PPD : 94.5

Step III: Reduction

V 1328

Check tool.

Check all valves.

Check all accessories.

Charge

2285 L water. Heat to 90°C with live steam and adjust volume to

2900 L at 90°C.

Charge

290 Kgs Iron Filings or Powder and

33-80 Kgs of H_2SO_4 80% W/W, in 10-15 minutes.

Stir at 90-95°C for ½ hour.

Charge wet cake of half batch of

		4-4′-nitroamino diphenyl-2-Sulfonic acid
(6 to 10 Hrs)		in 2-2.5 hours at 99-100°C maintaining pH at 5.5 to 6.0 (1/2 batch of w. c. obtained from 400 Kgs real PNA)
		Charge another
145	Kgs	Iron filings and
14.5 to 50	Kgs	of H_2SO_4 80%w/w.
		Test of soluble iron should be positive.
		Charge the remaining ½ batch wet cake of 4-4′-nitroamino diphenylamine-2-Sulfonic Acid in 2-2.5 hours at 99-100°C.
		After completion of Nitro wet cake charging check the test of soluble iron and pH. (It is 5.5 to 6.0) and report them. If the test is negative, charge corrective iron filings and sulfuric Acid 80% w/w/ Check the volume and report. (4000 to 4500 L)
		Close the manhole and reflux the batch for 5 hours at 90-100°C.
		Cool the batch to 90°C.
		Send a sample to Lab to check the completion of reduction. If reduction is reported incomplete, charge
57 to 70	Kgs	Iron filings and reflux for 2 hours more.
		Check the test of reduction again. If the reduction test is O. K. cool the batch to 90°C.
		Charge
235-260	Kgs	Soda Ash powder in 1 hour.
		Check pH. (It should be 8.0 ± 0.2)
4000-4500	Ltrs	Check volume and report. Charge
12	Kgs	Sodium Sulfite in 20 minutes.
		Blow the reduction mass to preset and preheated Filter Press FP 1001 and clarified the solution. Take 1450 Liters of water at 90°C and blow it to FP 1001 as 1st wash.
		Again take 1450 Ltrs of water and heat it to 90°C. Blow to FP 1001 as second wash.
		(V 1328-Time Cycle 22 to 24 hours.)

FP 1001: Filtration:

Check filter Press well.

Preheat the F press

Receive reduction mass from V 1328. Clarify the main batch liquor and collect the clear clarified liquor in V 1321.

Receive 1st wash at 90°C from V 1328.

Clarify thro' Press and receive the filtrate in V-1321.

Receive 2nd wash at 90°C from V 1328 clarify and pump the filtrate to V 1321.

Air the Filter Press well

Dump the Press.

Weigh the sludge and analyse it.

Weight of Iron Sludge obtained: 600-800 Kgs.

% N. V.	:	1.0-1.5%
Loss of product, Real Kgs	:	8 to 10.
Time Cycle	:	10-14 hours.

V 1321 10,000£ MSRL

Check Tool.

Check all valves.

Check all accessories.

Receive the clarified Amino liquor as Sodium Salt solution (main liquor) from FP 1001 into V-1321.

Check and report the volume and temperature

4000 L.

Receive 1st Wash. Note volume

5400 L.

Receive 2nd wash. Note volume.

6700-6900 L.

(The volume will be about 6500 Ltrs)

Analyse a sample of the solution by a Nitrite titre for real content. (Expected real: 772 Kgs).

Heat the amino solution to 80-86°C.

Charge

527 to 700 kgs 30% w/w HCl to get pH 0.8 to 1.0.

check and report pH and volume. The product dissolves as the Dihydrochloride

Add to it

10.0 Kgs Darco.

Blow the batch to precoated and preheated PRN 1025 and clarify a 80°C.

Wash the Nutsche sludge with

450	L	of hot water of 90°C, duly acidified with the addition of
34.0	Kgs	of 30% w/w HCl.
		This is an acid wash of pH: 1.0-1.5.
		Suck the sludge in the Nutsche well.
		(V-1321-Time Cycle 22 to 24 hours).
		PRN 1025 -FP 40″-20 fames
		Check PRN.
		Precoat PRN with
6.0	Kgs	of Hyflosupercel slurried with 20 L water.
		Preheat the PRN.
		Receive Amino Hydrochloride solution from V 1321.
		Likewise, receive acidic wash and clarify it.
		Air dry the sludge (To dump sludge) after 4-5 batches.
		(After 4 batches) 100-150 Kgs.
		(PRN 1025-Time Cycle 10-14 hrs.)
		V 1322
		Check tool.
		Check all valves.
		Check all accessories.
		Receive the clarified Amino Hydrochloride solution and the acidic Nutsche washings.
		Note volume and temp.
		Volume of the clear batch solution will be about
7000 ± 500	Ltrs	at 75-80°C
		Check pH.
		Heat to 80°C.
		While stirring, charge as required, about
200-250	Kgs	Soda Ash powder during about 3-4 hrs to adjust the batch pH between 6.0-6.3
		Stir for half-an-hour and confirm the pH to be strictly between 6.0-6.3
		Add
4	Kgs	Sodium Bisulfite.
		The final product precipitates out at this stage.
		Check pH; It should be 6.0-6.3.
		It not, adjust pH by adding Soda Ash or HCl in this range externally to 30°C.

Stir for 1 hour at 30°C. Filter on PN 1003.

(V 1322: Time Cycle 24 to 28 hrs.)

PN 1003/PN with stirrer 1004 Filtration:

Check the product nutsche. Receive batch from 1322.

Suck M. L. in batchegg V 1704. Wash W. C. with 6 × 125 Ltrs of cold water of 10°C.

Suck Dry the Nutsche cake. Dump the product Nutsche cakes and weigh.

Weight of wet cakes: Approx 900-950 Kgs.

Yield: Expected 648 Kgs. Real, 70.0-75%.

2. Reactions:

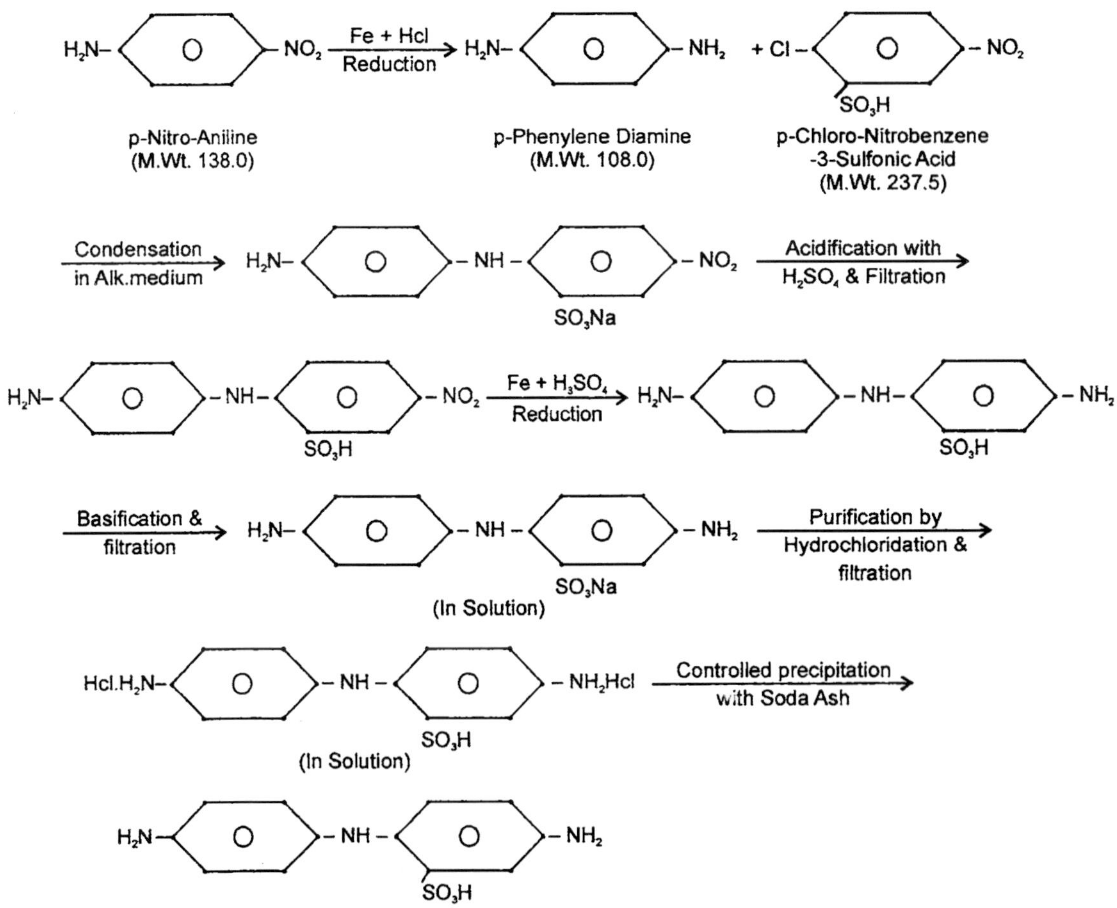

4:4′-Diamino-Diphenylamine-2′-Sulfonic Acid.
(M.Wt. 279.0)

33

Manufacture of Benzidine Di-Hydrochloride

		Charge in the Kettle
2000.0	L	of Methanol. Stir and charge
650.0	Kgs	of Caustic Soda, flakes.
		Add to it about
65.0	Kgs	of Black Catalyst.

Preparation of Black Catalyst:

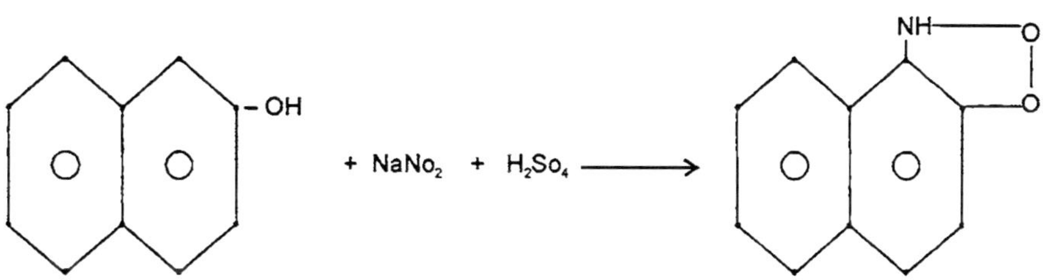

Black Catalyst
(80%)

Beta-Naphthol is nitrosated with $NaNO_2$ & H_2SO_4.

The yellw-coloured Nitroso-Beta-Naphthol is filtered out.

The wet cakes are added in 50% Sulfuric Acid solution containing some iron powder and stirred for a while.

It is then filtered and the wet cakes are used as such as Black Catalyst, without removing iron.

Then add during seven hours

1300.0 Kgs of Nitrobenzene.

The temperature at the end of addition is 91-92°C.

Then close the Audco Valve on the Kettle to completely stop the reflux.

Continue the reaction for six hours at 125°C and under a pressure of 60-65 p. s. i.

Most of the Hydrazobenzene formation is over, here itself. Then add only

75.0 Kgs of as is 40% w/w Formaldehyde solution (in ½ hour) and continue the residual reaction at the same temperature and pressure for Two hours.

The reduction to Hydrobenzene will be now complete.

Test for unreduced Azobenzene as follows:

Test : Take about 10 ml of the batch sample in a test tube. Add to it distilled water, shake well and decant out the water. Repeat this for 3 to 4 times. Now to the solid material remaining in the test tube add boiling distilled water and stir with a rod. If necessary heat the test tube over a burner's flame to boil.

Unreduced Azobenzene melts in hot water and the yellow molten material floats on the water surface. This denotes that unreduced Azobenzene still remains in the batch.

The reduction of Nitrobenzene to Hydrazobenzene is now complete and the Azobenzene test is negative.

Then the pressure is gradually released and

300.0 L of Methanol is distilled out at 64-66 degC by heating the Kettle with jacket steam as is necessary. Then add to the batch

300.0 L of water and distil out the residual Methanol along with water.

Note : 900 L of Methanol is actually used up in the reaction. About 600 L of Methanol is lost.

The Hydrazobenzene is then finally filtered out and washed.

The Hydrazobenzene is cream in colour. M. Pt. is 123-126°C.

Process of Benzidine mfre contd.

Note : For the manufacture of o-T Base, the pressure reduction does not work. It gives inferior quality of o-T Base. Thus for o-T Base, the reduction has always to be conducted under reflux at atmospheric pressure.

Benzidine Rearrangement

Charge in a Wooden Vat

2000.0 Kgs of as is 30% Hydrochloric Acid.

Stir and charge in it

1125.0 Kgs of broken Ice (15 slabs).

Note the temp., it about 10°C or lower.

Then charge in it the total quantity of the Hydrazobenzene wet cakes obtained under the reduction step during about 2 to 3 hours, holding the temperature of the batch below 25°C. Add more crushed ice, if required.

At the end of Hydrazobenzene charging, charge in the batch an additional quantity of

-2000.0 Kgs of as is 30% Hyrochloric Acid, allowing the batch temperature not to exceed 25°C by adding more crushed ice as required.

About

525.0 Kgs of additional quantity of Ice (7 slabs) will be required. At the end of charging the second lot of 30% HCl, determine the acidity in the bath as follows:

Test : Draw a sample of the batch slurry and filter it through a buchner funnel. Titrate 10 ml of this clear filtrate against a Standard 1N Caustic Soda solution to just no test on C. R. papers. The total acidity estimated this way should be between 12-13% w/vol as real Hydrochloric Acid on Batch slurry.

If not, correct it to this limit.

For 1 Mole of real Benzidine, about 8 Moles of real HCl are employed.

Then stir the batch without temperature-control for 10 hours. The temperature of batch is allowed to rise of its own during the stirring period.

Test the completion of Rearrangement as follows:

Test : Draw a sample of the batch slurry and filter it through a Buchner funnel. Suck the cake.

Dissolve 1 gm of the Benidine Di. HCl wet cake in 100 ml of hot water of 80-90°C. It must dissolve completely.

At the end of completion of the Rearrangement and Hydrochloridation, filter the batch through a Nutsche Filter at room temperature or at the final temperature of the batch. Suck the Wet Cakes.

Dump the Nutsche.

Centrifuge the Nutsche Cakes and wash with 30% Common Salt solution acidified to Congo faint blue test with HCl.

Dump the centrifuged cakes and weigh.

Record yield.

Weight of Benzidien Dihydrochloride wet cakes,

= About 1400-1500 Kgs

= 650 Kgs of real Benzidine (M. Wt. 184.0)

34

Manufacture of Meta Toluidine

In Cast Iron Vessel
Charge
350 Litres water and
then charge under stirring,
670 Kgs Iron Filings.
Close the manhole and add
11 Kgs Formic Acid.
Then heat to 100°C
In about 1½ to 2 hours
Using 35 lbs steam in the jacket.
Reflux Starts at 100°C,
then slowly in two hours
add at 100°C i.e. reflux temp.
250 Kgs Meta Nitro Toluene.
Stop steam in jacket and
Add
6 Kgs Formic Acid
Again start steam in jacket
& start reflux then slowly
in two hours
Add
250 Kgs Meta Nito Toluene
When addition of Nitro Compound
is over, Add

6	Kgs	Formic Acid
		Then reflux for 4 hours more
		Stop steam in jacket and
		Cool to 90°C
		Check completion of Reduction.
		Volume will be
		About 950 litres to 1000 litres.
		If reduction is complete
		Neutralize the batch using about
25	Kgs	Soda Ash-Strong test on B.Y.
		Before neutralizing the batch
		Adjust volume to
1200	litres	Again close the manhole and heat to 100°C
		Reflux for 15 mts.
		And start distilling oil
		Through the separator.
		Collect oil in a drum and let water to go back to Vessel.
		Distillation time-32 to 36 hrs.
		Distilled Oil-350 to 360 Kgs.
		When distillation is over
		Filter residue on F. P.
		Dump and discard.
		In Concial Vessel
		Charge this oil
		Stir for 30 mts and
		Settle for 2 hrs.
		Drain bottom water layer
		and collect pure oil in a drum
		Wt of Pure Oil
		328 to 335 kgs
		of about 98.5 to 99.% Purity
		Sp. Gr. of oil is 0.9.

Equipments Used

1. Cast Iron vessel of 2500 litres
2. M. S. Measure Vessel of 500 litres for MNT
3 S. S. Pot of 25 litres fir Formic Acid
4. S. S. condenser of 5 M^2
5. M. S. Seperator for collecting oil + water
6. M. s. Conical vessel of 500 litres for separating oil

Services

1. Cooling water in jacket of the vessel
2. Colling water in jacket of the condenser
3. 35 lbs steam in the jacket of the vessel.
4. Air for blowing the residual material to F. P.

Equipments

1. M/s or S/S Vessel for Sodium Methoxylate Solution
2. M/S Autoclave for Methoxylation Step.
3. M/S or S/S Conical Vessel for Isolation of oil and washing of oil with bottom outlet of the vessel.
4. Methanol Recovery Unit.
5. M/S Vessel for NaSH & Clarifier
6. M/S or S/S Vessel for Reduction Step
7. Cast Iron Vessel for Reduction Step using Iron-Formic Acid
8. Vac-distillation Unit.
9. Condenser of 5 M^2 and Seperator

Service

1. Cooling Water
2. Chilled Water
3. 35 lbs steam-Reduction Step
4. 100 lbs Steam for Under Pressure reaction and Vac. distillation

35

Manufacture of 2-Naphthylamine 4:8-Disulfonic Acid (Cassella Acid)–"C-Acid"

Sulfonator:

		Start cooling in jacket and check tools.
		Charge
410	Ltrs	98% H_2SO_4 from M. V. V102
		Cool to 15-16°C.
		Charge
53	Ltrs	65% Oleum from M. V. 101 at 15-17°C
		Stir and cool to 7-8°C by chilled water/Brine and send sample. Check volume.
		Charge
500	Kgs	Naphthalene at 5-10°C, slowly in 4 hours (24 hrs.)
		Drain jacket.
		Close manhole and side flanges.
		Stir for 15 minutes. (Keep vent open)
		Check volume and sample to Lab. Heat slowly to 20°C.
		Stir at 20°C for 15 min.
		Charge
919	Ltrs	65% Oleum from M. V. 1010 in 8-10 hours keeping temp 28-29°C (Record temp and rate of addition hourly).
		Let temp rise to 30-35°C.
		Stir at this temp for 8 hours.
		Check volume and send sample to lab.

Cool to 15-20°C.

Send sample to lab and check volume.

Blow sulfonated mass at 15-20°C to nitrator. AV 102.

Nitrator-mixed Acid Preparation:

		Charge
270	Ltrs	98% H_2SO_4 from M. V.V 102
		Cool to 20°C.
		Under stirring, start adding 98% HNO_3 from M. V.
160	Ltrs	
		Maintain temp below 30°C.
		Send sample for total acidity & Nitric Acid.
		Pump to mixed acid metering tank.

Nitration:

Receive Sulfonated mass from Sulfonator under stirring at 15-20°C.

Check volume. Cool at 15°C

Stir at 15-20°C for 15 min. Drain Jacket.

Heat slowly to 30-35°C

Stir at 30-35°C for 8 hours.

Send sample for % acidity and nitro content.

Blow to Neutralisation Vessel. AV 104

Lime slurry preparation vessel AV 303

		Receive water/Back wash from case.
		Make up volume to
1700	Ltrs	
		Under fast stirring,
		Charge
2800	Kgs	Lime stone powder.
		Transfer to....

Neutralisation vessel: (AV 304)

		Check tools.
		Take water/Back wash from Blow case
2300	Ltrs	

Close all valves and open valve to scrubber.

Receive nitro mass from 4-5 hours.

Stir 1 hour. Receive lime slurry from vessel over 6-7 hours (Control foaming)

Check pH, add lime-stone powder/hydrated lime if necessary, to adjust pH to 5.5.

Stir a 5.5 for 1 hour. Check pH again.

It should be 5.5.

If necessary, correct it.

Heat to 85°C.

Stir at 85°C for 1 hour. Blow to GN 8/9/10.

Filter Nutsche GN 8/9/10

		Receive neutralized slurry from AV 304
		Suck M. L. in V. 307 A/B. Give washings to the gypsum wet cake with
3000	Ltrs	Water (Temp 80-85°C)
		Collect M. L. and washings in vessel A/B V 107
		Total volume should be
75-7800	Ltrs	
		Nitro Solution (AV 305)
		Receive nitro solution from V 107 A/B Send sample to Lab for nitro content.
		Blow to AV 106 Reduction Vessel.

Reduction Vessel AV 306.

		Charge
1700	Ltr	water/Back wash.
		Start stirrer
		Charge
		(1100 Kgs) Iron powder. (Rust free)
		Add
40	Kgs	Acetic Acid. (or 84 L 40% H_2SO_4)
		Heat to 95°C. Stirr at 95°C for 1 hour.
		Add nitro solution uniformely in 6-8 hours, maintaining temp at 95°C.
		Send sample to lab to Check for completion of reduction.

		If complete charge 60 Kgs Soda Ash in abut 1 hour to get pH of 9.0.
		Blow to Filter Press. ½

Filter Press (FP 1-2)

		Receive from AV 306
		Give hot water washes and adjust volume in AV 107.
		Dump Air the Press. Reset and remove sludge.

Amino solution holding vessel

		Receive filtrate from FP ½
		Stir and send sample for analysis. Drain to PN.
8000	Ltrs	Blow to AV 108 via evaporator.

Isolation Vessel (AV 111)

		Receive Amino solution (Conc.) from AV 108
8000	Ltrs	
		Add
800	Kgs	Salt.
		Add.
325	Kgs	30% of HCl to get pH of 1.5.
		Cool to R. M.
		Drain to P.N

Yield

450	Kgs	C Acid.

36

Manufacture of N-Dioxyethyl-m-Toluidine

Tool	S. S. or Glass lined Kettle with jacket and anchor type stirrer.	
		Check tool.
		Charge
214	Kgs	m-Toluidine Oil (Sucked by vacuum)
		Stir.
		Heat to 80-85°C.
		Flush the unit with Nitrogen Gas for 5-10 minutes.
		And check that there is no leakage.
		If O. K., start passing Ethylene Oxide gas
		At 80-85°C till the reaction is complete.
		Ethylene Oxide gas cylinder is kept vertical on the balance.
		The rate of passing the gas is kept at 2.3 lbs pressure on the pressure gauge kept on the line between Kerosene-carboy scrubber and the Reaction Kettle, which indicates 2.5-3.0 Kgs per hour.
		It may be more in the beginning and less at the end. The cylinder is warmed at intervals by pouring warm water on it.
		Pass
180	Kgs	gas.
		Sample for testing when the exact theoretical amount of gas is passed, for setting point and charmatogram.
		Setting Point Less
		Monocompound + Ve
		Pass
1	Kg	more gas and retest.

Setting Point : 63°C.

Chromatogram : Comparable with Standard.

Flush with Nitrogen when the reaction is over.

Dump the material in S. S. trays at 80-85°C.

Cool in room temperature.

Material Solidifies into hard cakes.

Dump by cutting into pieces and weigh.

Quantity : 380 kgs.

Specifications:

Appearance	Slightly brownish yellow hard lumps.
C. V. (M. Wt. 195)	97.5%.
Matter insoluble in Dil HCl	Trace.
Setting Point-Dry Material	63°C.
Impurities by Chromatorgram on 100% basis	m-Toluidine absent.
Unknown impurities	Combparable to standard.
Red Rf 0.37)	4 × 5
Red Rf 0.60)	
Other impurities	more

Chromatognaphic Evaluation of N-Dioxyethyl-m-Toluidine

Solution of Product	2% (on 100% Basis) in Methanol
Application	5 µl = 100 µg.
Test Substance	m-Toluidine
Thin layer	ACME
Developer	85 Benzene: 15 Ethanol (Acetic Acid Atmosphere)

Time	1 hour.
Temp	Room Temp.
Detection	PNA Diazo N/80
Evaluation	Semi-quantitative.

1.0		
0.9		
0.8		
0.7		
0.6	0	Yellowsh Red
0.5	0	Red
0.4	0	Red
0.3	0	Red
0.2	0	Red
0.1		
0.0	Sample	Test substance

37

Manufacture of Armstrong Acid (Naphthalene-1:5-Disulfonic Acid)

1. Principle:

Naphthalene is sulfonated at 43-45°C for 8 hour with 65% Oleum.

2. Reactions:

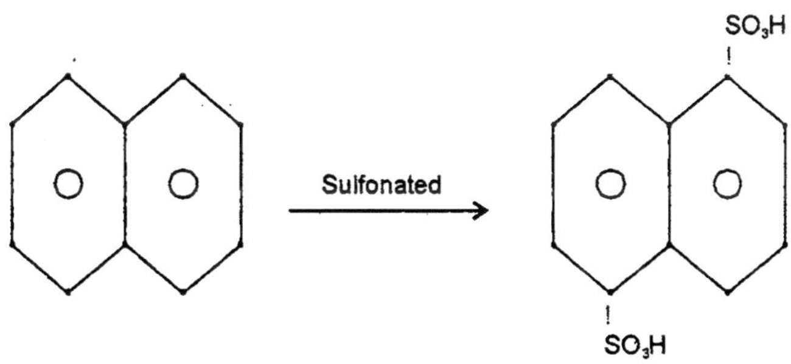

Naphthalene

Amstrong Acid (M.Wt. 288)

3. Process-Procedure:

Tool : V 100 2500 L Reaction Kettle with MS jacket, HS Anchor type stirrer, blow-leg SS removable.

Check tool, valves and all accessories.

Charge

960 Kgs as is 98% H_2SO_4 (475 L).

Stir.

Cool to 18°C with chilled water in jacket. Charge, with sieving under fast stirring.

396 Kgs ground Naphthalene over 4 hours,
maintaining temp at 20-22°C.

Record volume; it will be about 840 L.

Stir at 20-22°C for 2 hours.

Then charge, over a period of 8 hrs.

585 Kgs as is 65% Oleum, maintaining temperature between 43-45°C.

Stir for 8 hours at 43-45°C.

Record volume; it will be about 1300 L.

Check the reaction for completion of sulfonation. If O. K. blow to Drowning vessel.

Drowning:

Tool V-432. 12000 L MSRTL Reaction Vessel, MS rack-type agitator, MS blow-leg, 15 H. P. motor with variable speed.

Check tool, valves and all accessories.

Then charge

1800 L water.

Stir and add

1050 Kgs ice.

Volume will be about 2850 L.

Now receive the sulfonaed mass during 1-1/2 hours maintaining temp between 55-60°C. Do not allow to rise the temp above 60°C.

The solution is clear at 60°C. Record volume it will be 4100 L.

Check the clarity of solution.

Then charge, in one hour

530 Kgs as is Common Salt.

Stir at low-speed-25 RPM.

Cool to 25°C during 5-6 hours.

Stop stirring and all to stand for 12 hors at 20-25°C.

Check the Be of the filtrate. It should be 28-30° Be, and check the completion of precipitation.

If O. K. blow to Product Nutsche.

Filtration

Product Nutsche: 7.4 m^2 area, Tile lining carbon and ceramic.

Blow to PN, Suck M. L.

Wash the cakes with

900 L 15% w/v Brine in two lots.

Suck the cakes well.

Dump and centrifuge the cakes.

4. Yield:

Weight of Wet Cakes = Approx 657 Kgs.

= 450 Kgs real (M. W. 288).

= 50.5% Theory.

Sulfonated mass:

Analysis

(1) Solution clear.

(2) Insolubles absent.

(3) No smell of Naphthalene